The Organization of
Arab Petroleum
Exporting Countries

Recent Titles in
Contributions in Economics and Economic History
Series Editor: Robert Sobel

The Politics of Organizational Change
Iain Mangham

Change Agents at Work
Richard N. Ottaway, editor

Electricity for Rural America: The Fight for the REA
D. Clayton Brown

Commodities, Finance and Trade: Issues in North South Negotiations
Arjun Sengupta, editor

The Modern Stentors: Radio Broadcasters and the Federal Government, 1920-1934
Philip T. Rosen

J. Edgar Thomson: Master of the Pennsylvania
James A. Ward

Beyond the Adirondacks: The Story of St. Regis Paper Company
Eleanor Amigo and Mark Neuffer

Taxation of American Railroads: A Policy Analysis
Dennis L. Thompson

Indians, Bureaucrats, and Land: The Dawes Act and the Decline of Indian Farming
Leonard A. Carlson

Who Owns the Wildlife?: The Political Economy of Conservation in Nineteenth-Century America
James A. Tober

The Rise of the American Electrochemicals Industry, 1880-1910: Studies in the American Technological Environment
Martha Moore Trescott

Southern Workers and Their Unions, 1880-1975: Selected Papers, The Second Southern Labor History Conference, 1978
Merl E. Reed, Leslie S. Hough, and Gary M Fink, editors

The Organization of Arab Petroleum Exporting Countries

HISTORY, POLICIES, AND PROSPECTS

Mary Ann Tétreault

CONTRIBUTIONS IN ECONOMICS AND ECONOMIC HISTORY, NUMBER 40

Greenwood Press
Westport, Connecticut • London, England

Library of Congress Cataloging in Publication Data

Tétreault, Mary Ann, 1942-
    The Organization of Arab Petroleum Exporting
Countries.

    (Contributions in economics and economic history;
no. 40 ISSN 0084-9235)
    Bibliography: p.
    Includes index.
    1. Organization of Arab Petroleum Exporting
Countries.   I.   Title.
HD9578.A55T47   341.7'5472282'0601   80-24722
ISBN 0-313-22558-3 (lib. bdg.)

Library of Congress Catalog Card Number: 80-24722
ISBN: 0-313-22558-3
ISSN: 0084-9235

First published in 1981

Greenwood Press
A division of Congressional Information Service, Inc.
88 Post Road West, Westport, Connecticut 06881

Printed in the United States of America

10 9 8 7 6 5 4 3 2 1

This book is dedicated to
the memory of
Florence Kelley
and to
Rolland Richard Tétreault

Contents

Illustrations		ix
Tables		xi
Acknowledgments		xiii
Acronyms and Abbreviations		xv
1	Introduction	5
2	International Organizations and Arab Petroleum Policy	37
3	OAPEC: How It Works and What It Does	58
4	OAPEC and Arab Community	88
5	The Role of the Arab Nations in OPEC Bargaining	125
6	Prospects for the Eighties	168
	Appendix 1 - Agreement of the Organization of Arab Petroleum Exporting Countries	173
	Appendix 2 - Empirical Evidence of Arab Community	184
	Bibliography	201
	Index	211

Illustrations

Map

The Arab World   2

Figures

1 The Classification of Goods According to Their Nature in
   Consumption   19

2 The Relationship Among OAPEC, OPEC, and the Arab
   League   59

3 A Typology of Policies According to Their Nature in
   Consumption, with the Arenas Associated with Policy
   Formation in Each Policy Category   134

4 Estimated Crude Oil Reserves in Selected OPEC
   Countries, 1973-1979   138

5 Crude Oil Production in Selected OPEC Countries, 1973-1979   139

6 Public Policy According to Lowi and Wilson   162

7 Oil Production in Selected OPEC Countries During Two-Tier
   System in 1977   164

Tables

1 Mechanisms of Cartel Action Used by International Raw Materials Cartels Between the First and Second World Wars   25

2 Payments to Host Governments in Cents/Barrel of Exports   41

3 The OAPEC Joint Investment Projects, 1972-1977   72

4 A Partial Listing of OAPEC Trainees, 1974-1979   77

5 OAPEC-Sponsored Seminars and Conferences, 1974-1980   79

6 Percentage of the Value of Arab Country Imports and Exports by Various Countries and Regions for the Most Recent Year   93

7 Location of the Palestinians, 1969-1970   98

8 Location of Egyptians Abroad, 1975-1977   100

9 Arab Oil Tanker Fleets   106

10 Arab Gas Carriers   107

11 OAPEC'S Administrative Budget   110

12 Oil Revenues of OAPEC States in Millions of U.S. Dollars   111

13 New Arab Funds and Development Banks, 1973-1974   112

14 Member Equity Shares of Subscribed Capital   113

15 OAPEC Project Siting   115

16 Total Net Flows from OAPEC Members to Developing Countries, 1973-1976   116

17 OAPEC Oil Reserves   140

18 OAPEC Oil Production   140

19 R/P Ratios for Selected Members of OPEC   141

20 OPEC Oil Production Capacity, 1979   142

21 Excess Oil Production Capacity, 1979   143

22 Attributes of OAPEC Members   145

23 OPEC Confrontations, 1973-1977   146

24 Spot Market Premiums for Selected OPEC Crudes, January 1979   149

25 Official OPEC Crude Oil Sales Prices, End December 1978, Beginning January 1980   153

26 Oil Exports and Revenues   155

27 Distribution of the Rent From Oil   167

28 Results of Difference of Means Tests   185

29 Clusters Based on Religious and Ethnic Composition   189
30 Clusters Based on Indicators of Economic Development   189
31 Clusters Based on Trade Dependence   190
32 Clusters Based on Militarization   192
33 Pearson's r for Consumer Price Index   195
34 Pearson's r for Ln (Money Supply)   196
35 Loadings on Money Supply Factors   198

Acknowledgments

I have had a great deal of help in the preparation of this manuscript, and I am very grateful for all of it. Research materials were made available to me at the Shell Oil Company Library in Houston by Library Director Marilyn Johnson. The Old Dominion University Library, urged by my colleague John Ramsey, has begun a collection of books and journals which I have used to good effect. Richard Ashley arranged for my visits to the Klinesmid Library at the University of Southern California in March 1980.

I am particularly grateful to my friends Mary and Jim Ottaway, who housed me in New York and importuned their friends for me for access to critically important sources. James Riordan, treasurer of Mobil Oil, authorized his secretary, Florence Gosz, to arrange for my almost daily visits to the company's library during my stay in New York. All the resources of the Mobil Oil Library were available to me through the cooperation of Dorothy Fraser, Secretariat Library. Another Ottaway friend arranged for my visit to the Morgan Guaranty Trust Library. Mary's father, James Hyde, introduced me to Walter J. Levy.

I have been working on this manuscript for three years. George Tomeh, former ambassador from Syria to the United Nations and former OAPEC consultant, made available to me many sources of information I could not otherwise have consulted. During his year in Wisconsin, Dr. Tomeh permitted me to call upon him frequently for information and encouragement, even when we disagreed on the interpretation of events. Kamel Refaey, now a professor at the University of Riyadh, taught me the little Arabic that I know at this writing, and he also collected and translated for me articles in Arabic language newspapers. Gabor Galantai, professor of government at the University of Puerto Rico, has helped me in the organization of this material. He also allowed me to read the paper by one of his students that is quoted in Chapter 1 of this book. Georgiana Stevens was helpful and encouraging when I first began my work on OAPEC. I have quoted her in this manuscript as well.

Charles Doran, Gordon Smith, Fred von der Mehden, and Lou Griffin of Rice University and Scott Harris, now at the Department of Defense, were also very helpful during the early stages of this manuscript's preparation. Jamail Hussani of the Kuwait Embassy in Washington and Bader Aldafa of the Qatar Embassy were kind enough to allow themselves to be interviewed. Azmi Tubbeh of the OAPEC Information Service answered most of my many letters filled with questions about the organization.

The final research and preparation of this manuscript was supported by an Old Dominion University Summer Research Grant and an Old Dominion University School of Arts and Letters Grant. Manuscript typing was done by Anne Grey, who spent the entire month of July immersed in joint venture companies and oil prices. The figures were prepared by Debbie Miller, in spite of camera breakdown and the long wait for the man from Richmond to come and fix things. My friend, Christine Drake, prepared the map at the front of the book.

I would also like to thank my family. My husband and children have not seen very much of me this past year, and Richard says that even when my body was home, my mind was somewhere else, probably in the Middle East. I plan to be home in mind and body for at least a week before I begin my next project.

Acronyms and Abbreviations

ADWC	Arab Drilling and Workover Company
AECC	Arab Engineering Consulting Company
AMPTC	Arab Maritime Petroleum Transportation Company
API	American Petroleum Institute
APICORP	Arab Petroleum Investments Corporation
APSC	Arab Petroleum Services Company
APTI	Arab Petroleum Training Institute
ARY	Arab Republic of Yemen
ASRY	Arab Shipbuilding and Repair Yard Company
CIA	Central Intelligence Agency (U.S.)
CIPEC	Conseil Intergouvernemental des Pays Exportateurs de Cuivre (Council of Copper Exporting Countries)
CPI	consumer price index
GNP	gross national product
IBA	International Bauxite Association
IEA	International Energy Agency
MBD	million barrels per day
MEES	*Middle East Economic Survey*
NLRB	National Labor Relations Board
NOCs	national oil companies
OAPEC	Organization of Arab Petroleum Exporting Countries
OECD	Organization for Economic Cooperation and Development
OGJ	*Oil and Gas Journal*
OPEC	Organization of Petroleum Exporting Countries
PDRY	People's Democratic Republic of Yemen
PE	*Petroleum Economist*
PLO	Palestine Liberation Organization
PPS	*Petroleum Press Service*
UAE	United Arab Emirates
UAR	United Arab Republic
VLCCs	very large crude carriers

The Organization of
Arab Petroleum
Exporting Countries

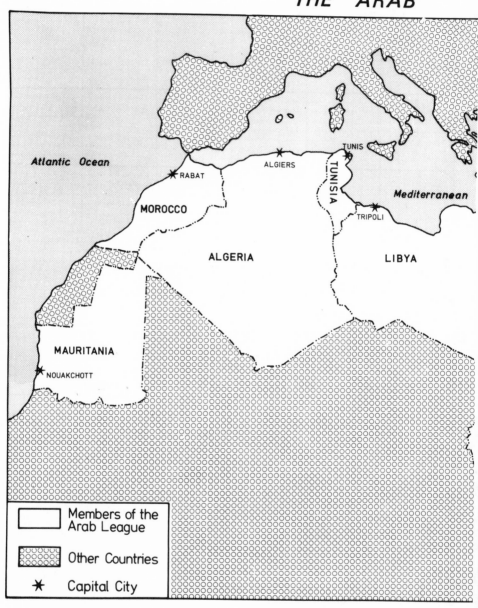

Atlantic Ocean

ALGIERS

TUNIS

RABAT

Mediterranean

MOROCCO

TUNISIA

TRIPOLI

ALGERIA

LIBYA

MAURITANIA

NOUAKCHOTT

Members of the
Arab League

Other Countries

★ Capital City

WORLD

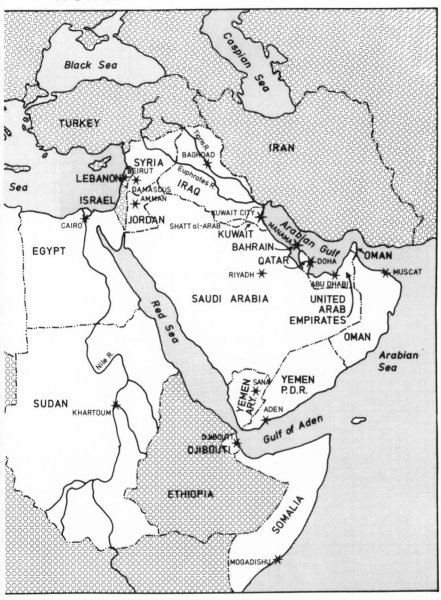

Chapter 1

Introduction

The Organization of Arab Petroleum Exporting Countries (OAPEC), was founded in January 1968. Little notice was paid to the new Arab organization by the American press or by scholars.[1] The *New York Times* index shows only two entries for OAPEC in 1968 and none at all in 1969. Few articles on this organization appeared in the *Times* until the fourth quarter of 1973, when the oil crisis prompted eight articles concerning OAPEC, the organization that had imposed production cutbacks in member-country petroleum production destined for export and that had embargoed entirely countries supporting Israel in the October 1973 war.

American oil industry publications such as *World Oil* and *Oil and Gas Journal* include a few references to OAPEC. *The Petroleum Economist* (formerly *Petroleum Press Service*) and *The Economist*, both British publications, have a somewhat better record than their American counterparts in their Middle East coverage generally. Even so, Westerners interested in OAPEC have had to rely primarily on regional journals, the most important of which is *The Middle East Economic Survey (MEES)*, for their primary source material. This work, like many others on the Arab oil industry, relies heavily on *MEES* as a primary source.[2]

Before 1973 OAPEC was virtually unknown to most Westerners.[3] When the OAPEC Council of Ministers voted to cut oil production as a means of exerting pressure on countries aiding Israel in the Ramadan-Yom Kippur War, American policymakers and commentators alike often confused OAPEC with OPEC, the Organization of Petroleum Exporting Countries. OPEC is the interregional oil organization that used the opportunities created by war, embargo, and production cutbacks to quadruple oil prices in December 1973 from their level the previous January. As a result of the interplay between OPEC and OAPEC policies in the fall of 1973, some analysts blamed "the Arabs" for changing the power relationships among oil producers, oil consumers, and the multinational oil companies.[4]

All the Arab members of OPEC belong to OAPEC as well.[5] But the fact that the seven Arab members of OPEC constitute a numerical majority of its thirteen member countries is not enough of a reason to conclude that they control OPEC. This is one of the issues that shall be explored in this work. A related issue is concerned with the nature of OPEC. Is OPEC really a "cartel" as some American scholars refer to it in their work on OPEC and other commodity organizations such as the copper exporters' group, CIPEC? Arab scholars disagree with Westerners on the matter of OPEC's cartel status. The cartel issue will be treated later in this chapter.

Disentangling the OPEC-OAPEC-cartel relationship might appear at first to be no more than a minor analytical exercise. But the questions, what is a cartel, does an oil cartel currently exist, and do the Arabs control this alleged oil cartel, are not simple questions. Many journalists and some scholars begin by assuming that the last two questions are already answered and they proceed from that point. This has led to considerable confusion in the minds of general readers and, more importantly, of policymakers.[6]

In a letter to *The Wall Street Journal* published on 16 February 1979, S. Baron of Oradell, New Jersey, implies that Saudi Arabia can dictate world oil prices to the rest of OPEC. "It is ironic that the President's hard-fought effort last year in Congress to supply $2.5 billion of modern military aircraft to Saudi Arabia was rewarded by a 15% OPEC oil price increase. Whether by design or coincidence, this increase in oil prices will pay for these additional military supplies within two years..."[7] Although it is possible that this is a willful misinterpretation of events (S. Baron may deplore arms sales to all Arab countries because of her/his position with regard to Israel), such an assumption with regard to the Saudi role in OPEC is credible only to those unacquainted with the way OPEC really works.

What I propose to show in a subsequent chapter is that OPEC is not monolithic and that its Arab members do not constitute a controlling bloc within the organization. This is an important point for American policymakers to understand. American policy with regard to any individual Arab oil-exporting country, such as Saudi Arabia, ought not to be heavily influenced by OPEC's oil-pricing policy. This was generally the case with the Shah of Iran, who continued to receive extensive military support from the United States despite his position as a "price hawk" in OPEC until December 1977.[8] But American-Arab relations are viewed by both sides through the prism of Israel. Because of this, the precise nature of the Arab (and thus the OAPEC) contributions to OPEC policy will be explored in some detail.

Although the cartel aspects of OPEC, and perhaps of OAPEC as well, have preoccupied analysts of world oil, a study of OAPEC is equally interesting to policymakers and scholars concerned with economic development. OAPEC made its greatest impression on the West through its embargo and cutback decisions. Its greatest impact on the Middle East has been through its development projects, its efforts to promote the economic integration of the oil sector in the Arab countries of the region,[9] and its role in technology transfer. OAPEC is unique among international organizations, particularly among organizations of developing countries, because it has been able to allocate great wealth to capital projects. The reasons for this are twofold. First, as oil exporters, several OAPEC member countries found themselves with more money than they could spend on national development and welfare projects after the 1973 price increases for oil. These countries have enough money to undertake expensive international projects as well as their own capital projects.[10] Second, two social forces, Arab nationalism, a transnational phenomenon including political action and simple individual indentification with the idea of a single Arab nation, and Islam, always a potent force in Middle Eastern politics, work together to reinforce efforts at increasing Arab international community. Arab community, defined here as the proliferation of personal, political, and economic ties between the citizens and governments of the Arab states, is not a special preserve of OAPEC. There is a plethora of Arab and/or Islamic mutual aid organizations that were set up to foster regional interdependence. A major motivating force behind the establishment of these organizations was the desire to decrease Arab dependence on the nations of the developed West.[11] OAPEC is similar to these other organizations in its community intent.

OAPEC is also different from other Arab and Islamic organizations. First, it is a commodity organization. Second, it has chosen to sponsor capital investment in the oil industries of member states in a novel way: through a joint venture approach in which member governments are the shareholders and members of the corporate boards of directors. This aspect of OAPEC sets it apart from non-Arab international organizations as well. Thus the study of OAPEC is also of some interest to persons concerned with new kinds of international organizations per se.

The aim of this work is to provide as comprehensive a picture of OAPEC as possible. It begins with the fitting of OAPEC into historical, regional, and issue contexts: how does OAPEC relate to other international organizations that aim to set petroleum policy and that foster community among their members? Next, it focuses on specific policy areas: economic development, Arab community, and petroleum pricing, particularly the OPEC-OAPEC interaction. Throughout, the fact that OAPEC policymakers are also OPEC, Arab League, and governmental policymakers both explains the extent of interorganizational coordination and cooperation and confuses the issue of which specific body is responsible for which policies. It demonstrates the utility of institutions that may protect individual governments from the consequences of collective action. One recalls the former shah of Iran pushing oil prices higher and higher in 1973 while denying his involvement in the matter of the oil shortage that supported the price increases.[12]

The rest of this chapter will be used to define concepts of importance to the policy areas of major concern to OAPEC. These are "economic development and technology transfer," "Arab community," and "cartel."

Economic Development and Technology Transfer

Policymakers in OAPEC member countries are deeply involved in the problems of economic development, indeed, almost to the point of obsession. For most of these persons, economic development in their countries is an elusive goal, a possibility only, and not by any means the sure end of their efforts.[13] Some feel that their only opportunity to achieve economic develop-ment is the one created by the post-1973 oil revenue bonanza.[14] But in spite of the financial resources that most of them devote to paving the road to development, domestic economic policymaking for the OAPEC countries is often perceived as a race against the time when the oil and the money will run out.[15]

The ability of a national economy to be self-sufficient and to meet new conditions is one definition of economic development. Several models or recipes for economic development have been suggested, mostly by First World economists. At least one Arab country, Egypt, appears to have tried them all. Under Muhammad Ali, state-owned enterprises were created.[16] After the collapse of this experiment, Egypt reverted to an agricultural economy specializing in long staple cotton during the nineteenth-century free trade era. Import substitution industrialization began after World War I, and after the revolution of 1952, President Gamal Abdal Nasser repeated Muham-mad Ali's venture into state-owned industries.[17] Anwar el Sadat has put his faith in direct foreign investment and large infusions of foreign aid to achieve economic development.[18] So far, none of these has worked.

A variety of reasons has been suggested for the inability of Third World economies to achieve self-sufficiency. Chief among these are scarcity of capital and a shortage of trained personnel.[19] For half of the OAPEC member countries, capital shortage is not an issue. These nations have had enough money to spend on development and for greatly increased levels of consump-tion for most of the years since the oil revolution of 1973.[20] This means there has been capital available for infrastructural development and for industriali-zation. The limit to the use of this capital is the capacity of an economy to absorb it: ports to accommodate imports; roads to carry imports and domes-tic products to final destinations; people to work in agriculture, transporta-tion, and industries. However, the OAPEC countries that have excess capital are also short of native labor. Since the huge increase in their annual incomes, these countries have increasingly imported people as well as goods. Kuwait is estimated to have guest workers making up 60 percent of its population.[21] Qatar's guest workers are about 60 percent of its population.[22] Although the guest workers are necessary to operate the countries' economies, they are also potentially destabilizing politically. Guest workers do not receive the same social benefits as native populations, and some may resent their lower status in the countries where they work. The region has already experienced turmoil from non-native populations, which form a substantial percentage of many nations' educated populations. The conflict between Jordan and its large Palestinian minority in 1970–1971 is a constant reminder to the governments of Saudi Arabia and Kuwait, which also have large numbers of

Palestinian workers, and workers of other nationalities, in responsible positions.

If the OAPEC countries can overcome capital and labor constraints, their next problem is to put together a development strategy. As well as downstream operations in the oil industry itself, some OAPEC countries have chosen to enter energy intensive industries, such as aluminum and steel, and industries that use petroleum as a feedstock, such as petrochemicals. Although it seems logical for the near term to invest in industries that capitalize on the competitive advantage oil exporters have in energy, some question the advisability of concentrating so heavily in energy-dependent areas, an issue that will be discussed later.[23]

Kuwait is one country that has gone beyond oil and gas in the energy field to specialize in solar energy. However, much of its solar technology is purchased from developed countries rather than being developed in Kuwait.[24] The problem created by concentrating on foreign technology stems from the definition of economic development that was given in the beginning of this section: the ability to be self-sustaining. Even if Kuwait were to obtain virtually all of its power requirements, other than for transportation, from solar energy, the benefits would lie almost entirely in the area of conservation rather than in development because of Kuwait's dependence on outsiders for technology.

Arab efforts in solar energy are an example of purchased technology like the spread of "turn-key" plants in the region. A turn-key plant is one built by foreigners, perhaps by one of the international "instant economic development" corporations such as Bechtel or Fluor that install entire factories ready to begin operation in countries that can afford to buy them. A turn-key plant often signifies an import substitution strategy. The purchasing country wishes to produce something it has previously purchased from abroad. If a shortage of foreign exchange is a problem for the purchasing country, import substitution may be a good strategy to adopt. For the Arab oil exporters with huge dollar surpluses, import substitution as the chief strategy probably is not wise. Some of these countries have the unique opportunity to develop in a situation where money and therefore consumption and investment are only minimally limited over the near to medium term. They can afford to buy abroad much of what is needed for immediate consumption. They have the time to educate a generation of scholars and specialists to be the foundation for economic self-sufficiency in the next century. They can begin industrialization in areas where they have a comparative advantage, such as in energy industries, and then diversify into service industries over the next fifteen to twenty years.[25] Turn-key plants may have some utility if they are operated by native personnel, and gradually modified by native technologists. Solar energy research may be even more likely to pay off in terms of the development of a specifically Arab technology owing to a favorable climate and extensive government support. However, this payoff, if it is achieved at all, is still in the future.

A model for Arab economic development would take into consideration the unique financial position of Arab countries as members of the Third World. Industrialization in the Arab world should aim for self-sufficiency in the future by investing heavily in education and in institutions that promote research and development. At this time, most Arab universities concentrate more on teaching than on faculty research.[26] Opportunities for scientific research and exchanges of information should be increased, as well as opportunities to study abroad in First World and other Third World countries. The replacement of foreign personnel by native workers should be a top priority in Arab industries. Opportunities to get good jobs in their native countries might keep the most promising graduates at home instead of contributing to the brain drain to the developed West. Because Arab human resources are so scarce and because there are limits to the money, the Arab governments should plan together and coordinate their development policies instead of duplicating projects across the region.[27]

Finally, every effort should be made to encourage Arab governments and industries to patronize Arab industries, and to consult with the many Arab experts available to them when planning development projects. In many instances, Arab expertise is bypassed in favor of the importation of foreign techniques and personnel.[28] As a result, the turn-key plant, very much a feature of the Arab world, represents only a surface change in Arab economies: it produces something they did not produce before. For a developed country purchasing a turn-key facility, the new plant becomes part of an already established technical network.[29] It can be repaired, duplicated, or even improved upon by workers in the purchasing country. In the Arab world, where domestic expertise is bypassed in choosing, designing, and often in operating a turn-key facility, the technology it involves remains alien.

The issue of technology independence is crucial in the economic development of the Arab states. One reason OAPEC concentrates on downstream development in the oil industry is to create the opportunity for Arab scientists and workers to gain experience in a high technology industry.[30] But this decision involves many drawbacks. Oil-importing countries continue to build refining capacity, adding to the current oversupply and its exacerbation by new and increasing oil producer installations. Oil consumers prefer to buy products made in their own countries. These will probably continue to be cheaper for some time than products from producing countries and are more attractive to consumer countries because crude purchases require lower foreign payments than product purchases do. Most consumer countries have been unable to achieve a positive trade balance since oil prices quadrupled in 1974. They are unlikely to be interested in worsening their payment positions while increasing domestic unemployment of capital and labor by buying foreign products.

Another problem for OAPEC nations involved in downstream investment is its high cost and limited usefulness. Oil tankers, refineries, and petrochemical plants are expensive. They cost more for an OAPEC nation than they do for a developed country purchaser.[31] Once in place, they are dependent on continued petroleum production. For most OAPEC members, this means a useful lifetime of less than twenty years.[32] Thus, the gamble on downstream investment as a means to achieve economic development is very risky.

In assessing the role of OAPEC in regional economic development, attention will be paid both to the projects in which the organization participates and the contribution it makes to eventual technological independence for the region. The projects are interesting because of their novel corporate structure and because some might become very successful in creating employment and revenues for the member states. OAPEC's efforts at achieving technology independence are even more remarkable because this is where the organization is actually more of a pioneer than it is in development projects. Assessment of its success, however, will be difficult in both areas. OAPEC and its development efforts are too new to have succeeded or failed. At this point an evaluation of the contribution of OAPEC to regional economic development will have to rely on assessing its ability to elicit cooperation from its member states and on its efforts in laying a foundation for eventual technological independence.

Arab Community

Arab international community is unique in the world because of the existence of a high level of transnational identification among the Arab masses and, to a lesser degree, Arab political elites,[33] with an Arab nationalism transcending the boundaries of the individual Arab states. This transnational nationalism has been compared by historians to medieval Christendom.[34] Its unifying elements are a common, proslytizing religion and a common written language of cultural and theological importance.[35] Whether or not one regards this parallel with medieval Europe as accurate or helpful in understanding the modern Arabs, through their shared language and cultural and historical past, the Arab countries have a very different approach to nationalism than, for example, the countries of modern Europe.

Historically, the Arab people, with few exceptions, did not live in autonomous nation-states with individual traditions of statehood or notions of a collective past. In fact, during the last quarter of the nineteenth century most of them experienced the humiliation and impoverishment of colonial exploitation by the Ottomans, who were titular suzereins in most of the Middle East, and by Europeans, who occupied the Middle East through their businesses and financial institutions.[36] The nineteenth-century revival of interest in the Arabic language and the glories of the Arab past probably swept through the Middle East so intensely because of the low status of Arabs at that time. The revival of the Arabic language was symbolic in its recall of a time when Arabs ran their own affairs and conducted their business in their native tongue. It was also a substantive connection uniting the Arabs of that day through their vernacular use of a common language. Finally, the respect for Arabic as the language used by God to communicate with a chosen people also served to increase the perception of Arabs that they were a group with something special in common. This period of the "Arab awakening" gave the Arab peoples a collective past as Ottoman and then European imperialism gave many of them a collective present. Out of this new perception of shared past and present arose the idea that there should be an independent Arab nation.[37]

During the First World War, Arabs fought against their Turkish masters under assurances given by Great Britain to Sharif Husain of Mecca, titular religious leader of Islam, that the British would sponsor an independent Arab nation after the war.[38] But the British had also promised the French that they would share the Middle Eastern and North African empires of the Ottomans between themselves after the war. And separately, to European Zionists, the British promised a national home in Palestine, part of the territory that the Arabs wanted to reclaim for their independent nation and which they believed the British had promised to them.[39] When the war was over, the Arabs were denied their independence and a system of "mandates" or protectorates was set up under the auspices of the new League of Nations. This gave international sanction to the carving up of the old Ottoman Empire into spheres of influence for the French and the British.

Arab nationalism stayed alive under the mandate system because the states that the system set up were not cohesive nations but rather arbitrary subdivisions of the Fertile Crescent to suit British and French interests. The idea of an "Arab" nation was therefore more appealing to persons living in the artificial states than identification with their own new nations. In the words of Malcolm Kerr, "Pan-Arabism ... arose as the reflection of the national identity problems of Syrians, Lebanese, Palestinians and Iraqis...."[40] But the same mandate system also gave rise to particularistic nationalisms in the new states, especially in Palestine and Lebanon, which had large Christian communities that feared absorption into a unified Arab, and thus Muslim, state.[41] Britain and France each installed its own form of administration and education which promoted differences in these areas that the Ottoman administration had not. The French made great efforts to integrate the economies of their mandated territories with the metropolitan economy. This had the effect of isolating these countries' economies from those of the other states in the region.[42]

Finally, both the French, who were determined to rule directly in their mandates, and the British, whose main interest was to keep the way to India under control friendly to themselves, interfered with the local governments and eroded their legitimacy. The French were very direct. They did not hesitate to use military force in Iraq to suppress popular liberation movements.[43] In 1920, the French sent their army into Syria to drive Faisal, son of Sharif Husain, from the same throne they had installed him on just a few months before. The British were more subtle in installing rulers in their mandates who would be friendly toward British interests. They managed this in Iraq by expelling the popular native leader, Saiyid Talib, and then running a "plebescite," which resulted in the "election" of the Syrian Faisal to the Iraqi throne. A "two birds with one stone" policy, this action allowed the British to repay the family of Husain for its role in the war while getting a ruler in Iraq who would be dependent on the British to keep his throne.[44] Similarly, the British gave the east bank of the Jordan River to another son of Husain, Abdullah, who had stopped there with his army on his way to attack Syria. At that time there was neither a geopolitical nor an economic reason to create the state of Transjordan, and it could not have survived without massive British subsidies that also served to tie it more closely to British interests.[45]

Two Arab nations in the Middle East did not suffer from the national identity problems that kept the idea of Arab nationalism alive. One of these, Saudi Arabia, had not been geographically divided and did not have a colonial past. Apart from the western coastal area, the Hejaz, in which the holy cities Mecca and Medina are located, the kingdom was mostly empty desert, which discouraged colonial occupation.[46] The nomadic population was not easy to administer and because the people were so poor, no real effort to do so was made either by the Europeans or by the Ottomans. In the age of Ottoman imperialism, most of the Arabian peninsula remained a backwater of puritanical sectarianism. The present-day kingdom was united over a twenty-five year period by 'Abdul 'Aziz ibn Sa'ud, a descendent of the rulers of the eighteenth-century Wahhabi empire.[47] Beginning in 1902, he returned from political exile in Kuwait to retake his capital (Riyadh) and to unite the eastern portions of the kingdom through a comprehensive policy of military conquest and appeals to religious brotherhood.[48] Although Sa'ud managed to maintain good relations with the British in spite of his neutrality during World War I, the British would have preferred that Husain, already ruler of the Hejaz, become the king of Arabia. Instead, after a religious disagreement with Husain, Sa'ud conquered Mecca and Medina in the early 1920s and added the Hejaz to Saudi Arabia.

In contrast to the tribal social structure and geographical isolation of Saudi Arabia, Egypt lay at the crossroads of Asia and Africa. After the completion of the Suez Canal, Egypt provided the fastest and cheapest route between India and the Mediterranean. Increasing its attraction to an imperial power was the fact that, unlike Saudi Arabia, Egypt was the home of sedentary farmers relatively easy to tax and administer.[49] Egypt also had been a nation with relatively stable boundaries longer than any other twentieth-century Middle Eastern nation, and it had an independent historical past different from and as glorious as that of the Islamic Arabs.[50] Not surprisingly, from the beginning of the mandate period until the Second World War, Egypt was an unfavorable environment for movements aimed at creating a single Arab nation in spite of its colonial past. However, there had always been extensive social contact between Egyptians and other Middle Eastern Arabs; Egyptian governments had often taken an interest, sometimes an active one, in the affairs of the Fertile Crescent and Arabian Peninsula countries. After the Egyptian revolution of 1952, the country became something of a role model to other Arab nations interested in divesting themselves of autocratic traditional regimes. Especially after the establishment of the state of Israel on its borders, Egyptians saw themselves as Arabs and other Arabs regarded them in the same way. However, the Egyptian variant of Arab nationalism has been generally more positive than the ideologies rooted in the Fertile Crescent countries[51] because it does not challenge the authority of the nation state to the extent that Arab nationalism has done there. In addition, Egyptian Arab nationalism has always envisioned a large role for Egypt.[52]

After World War II, rejection of the colonial experience was still a large component of Arab nationalism.[53] The apparent lack of legitimacy of the governments of various Arab states (by this time including Egypt) increased the attractions of Arab nationalism.[54] A third factor promoting an Arab identity in the region as a whole was the establishment of the state of Israel. This third element also served to exacerbate the bitterness arising from colonialism and Arab disenchantment with their national governments. Israel was a constant reminder of European colonialism because it was (and is) widely viewed as a Western colony set up for the benefit of European Jewry. The perceived intention of the state of Israel to expand to the extent of the historical boundaries of the biblical Jewish nation meant that Israel itself had imperialistic designs in the region, an impression strengthened by its actions in subsequent Arab-Israeli wars and in the extension of Jewish settlements to occupied Arab lands. In addition, the ineffectiveness of the Arab states in the 1948 war aggravated the lack of legitimacy of the Fertile Crescent states' governments, particularly that of Syria.[55] Anti-Israel policies of the Arab League (such as the boycott against Israel and the admission of stateless Palestine to league membership) have been successful in keeping Arab, and specifically Palestinian Arab, grievances against Israel alive for Arabs and non-Arabs alike.[56] Similarly, the common hatred of Israel has helped keep Arab nationalism alive, even though the possibility of forming a single unified Arab nation has become increasingly remote since the 1940s.[57]

There were some attempts at partial Arab unification, which was permitted under the Charter of the Arab League. The late 1950s saw the peak of partial unification activity. Three unions were announced in 1958: the United Arab Republic (UAR), a serious attempt at unification between Egypt and Syria that lasted until 1961;[58] the United Arab States, more a statement of intent than actual unification between the UAR and the Kingdom of Yemen; and the Arab Union, which appeared to be no more than a political reply by two states that felt threatened by the formation of the UAR, Iraq and Jordan. The failure of the UAR was a serious blow to the hopes of Arab nationalists.[59] Since then, intentions to try partial unification or even limited partial unification[60] have been announced from time to time, but few of these plans have been implemented.

Arab nationalism was less important as a political movement in the 1970s. The existence of the Arab states as geographical entities for more than a generation has gradually given them a kind of legitimacy in the eyes of their citizens. And although colonialism was nearly universal, the fact that each Arab nation had its own liberation experience as part of breaking the ties to the mandate powers has given each something of a "nation-building" history, analogous to the exploits of Sa'ud. More recently, yet another "liberation" from colonialism was experienced by some of the Arab states. This liberation was from the multinational oil companies. Because the oil-exporting nations had been forced by company policy to forge individual relationships with the oil companies operating on their territories, liberation through the nationalization of state petroleum industries or through individual country-company participation agreements has also been a nation-building experience for each oil-exporting Arab state. Even Saudi Arabia, the Arab nation without a conventional colonial past, has had its own "liberation" from Aramco.

Another factor dampening Arab nationalism has been the changing position of Egypt with regard to the conflict with Israel. In 1973 Egypt attacked Israel on its own, consulting only with Saudi Arabia and not one of the Arab "confrontation states."[61] The half-hearted implementation of the OAPEC oil production cutback and embargo against the United States and other supporters of Israel by Iraq and Libya was justified by the governments of these countries as the result of their not having been consulted before the Egyptian attack.[62] When the Egyptian government became committed to reaching a peace agreement with Israel as the result of the stunning visit by President Sadat to the Israelis in November 1977, Egypt placed itself beyond the political pale from the point of view of many of the Arab countries. The pivotal geographical, cultural, and political position of Egypt in the Arab world makes its inclusion in any unity plan necessary to establish the credibility of the plan and to contribute toward its success. Egypt's peace treaty with Israel, more than any other factor or event in the Arab world, has broken the back of Arab nationalism as a means to political unification.

In spite of all this, Arab unity is still very important as an ingredient in the domestic politics of each Arab state.[63] And even though the possibility of there ever being a single Arab state becomes more remote daily, Arab peoples identify themselves as Arabs about as frequently as they identify themselves as Egyptians, Syrians, and so forth.[64] Arab community, possibly differently from other forms of international community, has its basis in history and in general mass support. This makes a formal investigation of inter-Arab political and economic ties a plausible undertaking despite the dearth of empirical evidence supporting a finding of international community in the region.[65]

International Community

A variety of models appears to underlie the international relations approach to international community. An example is the "security community" outlined by Karl Deutsch and his associates. This concept includes political "amalgamation," that is, the formation of a single, central government with the ultimate responsibility for political decisionmaking. It also includes the formation of a less tightly organized but still "integrated" system in which masses and elites share a sense of common interests and international institutions support a pattern of peaceful changes in relations among the participants.[66] Another community model is the "subsystem" or "region." Karl Kaiser has subdivided this model into three categories: the transnational society, where nongovernmental elites carry on the bulk of international transactions; the intergovernmental regional subsystem, involving cooperative behavior in a specific area among countries of a region, such as a military alliance; and a comprehensive regional subsystem, apparently comparable to the Deutsch et al. pluralistic security community, an international relationship that stops short of amalgamation or political unification, athough it is characterized by high levels of integration in other sectors.[67] Bruce Russett finds the boundaries of regions through the factor analysis of a wide range of political, geographic, and economic variables.[68] Charles Kegley and L. Howell use a similar procedure to find dimensions of integration in a predefined region.[69]

Underlying each of these models is a picture of community at the international level that includes shared values, geographical proximity, and behavioral indicators of international transactions such as trade and communications flows. Transactions are viewed both as means of effecting community and as evidence that it exists.[70] The usual unit of analysis in the studies mentioned, and others of the same type, is the nation-state. International communities are defined through the empirical analysis of behavioral indicators. Even scholars who concede an important role to international organizations, such as Deutsch and his associates or Roger Cobb and Charles Elder,[71] emphasize the behavioral bases of international community.

A second major approach to the study of international community has been that of international relations scholars whose unit of analysis is not the nation-state but rather the international regime created by international agreements, such as those establishing common markets, or regulatory mechanisms for international finance, such as the International Monetary Fund. Rather than seeking to define community boundaries or to classify various kinds of international community, the object of these studies has been, in the main, to discover the process mechanisms leading to international integration or the unification of one or more political or economic sectors common to all the participants.[72] These are not "community studies" in the sense of the Russett or Kaiser works. The community in these integration studies is a given fact. The search for process mechanisms is guided by the theoretical picture of community common to the earlier studies mentioned: transactions,[73] shared values,[74] and the regionalism implicit in the international agreements defining community boundaries. However, the integration studies place a great deal of emphasis on the role of the international political elites that manage the integration regime. The need of the regime itself to function in an appropriately comprehensive manner is thought to lead to the enlargement of the scope or the responsibility of the international organization.[75] National issue politicization or outright backsliding on the part of national elites are thought to decrease the relative power of the international organization.[76]

International integration studies appeared to have hit a natural stopping place even before the oil crisis of 1973 altered the emphasis in the study of international relations to include political and economic interdependence as a major explanator of interstate politics. Theoretically, integrationists had had difficulty in defining what integration was, a process, an end state, or some combination of these.[77] Because of the wide disparity among what each scholar meant by integration, it was difficult to view the various studies as a single body of research. Another difficulty was the seeming inability of Europe, the model case upon which the integration paradigm had been based, to achieve increasing levels of sectoral unification as the spillover theory had predicted.[78] Indeed, even in what were widely regarded as Europe's internationally integrated sectors, the integration was found to have astonishingly variant levels of enforcement in the several common market states.[79] International relations theorists such as Roger Hansen, Louis Cantori, and Stephen Spiegal quarreled with the functionalist model of integration with its spillover and spillback mechanisms because of its failure to include conflict as an explicit factor in interstate relations, even in cooperative relations,[80] and because of its assumption that economics and politics were a continuum in interstate relationships rather than different policy areas entirely.[81] Scholars who had been involved in the integration studies themselves criticized the narrow approach of some of their work and appeared actively to be seeking a new approach to the study of international relationships.[82]

The oil crisis of 1973 made it obvious that developed countries were not independent of the effects of decisions made by smaller, less developed nations. The new focus on international interdependence as the model of interstate relations for the 1970s relied heavily on earlier work by economist Richard Cooper.[83] Cooper assumed that integration would lead over time to increasing levels of interdependence or sensitivity of economic transactions between two or more nations to economic developments within those nations.[84] He divided interdependence into two types: structural interdependence, concerned with the international linkage caused by transnational behavior such as the demand for foreign goods, the international movement of labor and foreign investment decisions, and institutional interdependence, caused by the institution of joint decisionmaking regimes such as the European Economic Community, the World Bank, and the international financial agreements linking the economies of participating countries.[85] Although the Cooper categories were not adopted by political students of interdependence, Cooper's concept of sensitivity is the core around which more specifically political models of interdependence have been developed.[86]

As the interdependence model gained popularity among students of international relations, separation of policy areas into "high" and "low" politics became less and less acceptable. Cooper pointed out that interdependence would eventually "democratize" all of foreign policy by making matters which had formerly been the subject of elite-level "high politics" parts of policy packages including more traditionally domestic concerns such as trade policy.[87] Precisely this disorderly mixing of issues into bases for ad hoc policy coalitions is regarded by Ernst Haas as one of the factors making integration theories obsolete.[88]

In spite of the apparent mass movement of international organization theorists from integration to interdependence, integration studies themselves are not dead, nor is the theory associated with them. Integration as a concept has been adopted by students of development and modified to reflect the differences between developing areas and the European paradigm. Aaron Segal in the 1960s and Lynn Mytelka in the 1970s revised the traditional integration theory to describe economic integration in the Third World.[89] Their concentration on distributive and redistributive relationships among members of Third World common markets places their work among that of collective goods theorists[90] as well as that of regional integration scholars.

The concept of collective or social goods was developed by public finance economists to deal with instances of market failure.[91] If goods are classified according to their nature in consumption, the typology in Figure 1 can be created. Rival and nonrival refer to the characteristic of exhaustibility. If a good is rival in consumption, the amount an individual has decreases the pool of that good that others may consume. If we are sharing a box of popcorn, what I eat is unavailable to you. Exclusive and nonexclusive refer to the ability to deny feasibly, that is, without extraordinary effort, any individual from benefiting from a good once it is provided for anybody else. For example, once a neighborhood has installed street lights they light the way for strangers passing as well as for those who have paid to have them put in. Governmental intervention is thought to be necessary to provide nonexclusive goods in order to apportion benefits and costs across a population with some standard of equity as the guide. Otherwise, such goods would not be provided in socially optimal amounts.[92]

Figure 1: *The Classification of Goods According to Their Nature in Consumption*

	Rival	Nonrival
Exclusive	Private Goods	Public Utilities
Nonexclusive	Common Goods	Pure Public Goods

"Distributive" and "redistributive" as policy categories are associated with the work of Theodore Lowi.[93] Lowi says a distributive policy is one where governments give private goods to individuals. Examples are subsidies, land grants, and licenses. When governments create common entitlement programs funded through revenues collected from a noncoincident population, these are redistributive policies. Examples are social security, which taxes workers to pay retirees (among others), and public assistance programs, which tax property owners, workers, and consumers to pay the propertyless and unemployed.

Redistribution can occur automatically as a result of the formation of a common market among developing countries with variant levels of initial development.[94] Because industry prefers to settle in areas with greater levels of infrastructural development such as roads, banks, or a more highly educated labor force, the already stronger members of a common market group will expand economically at the expense of the weaker members. The expanded market benefits the larger industries of the stronger members and stifles the development of competitive industries in the weaker.[95] According to Lynn Mytelka, the only way for such an arrangement to survive is to allow each member to prosper, even if it means that comprehensive redistribution as a common market policy has to be instituted.[96] If redistributive measures are part of common market agreements from the start, it is less likely that their implementation will produce conflict in the organization. Mytelka does note in an ecological case that the degree of flexibility open to members in their bargaining over benefit allocation was directly related to the ability of the members to arrive at compromise solutions.[97]

Even though integration studies are still being done, a newer model for the study of cooperative international relations is economic interdependence and its extension to political behavior. As in the earlier studies of community discussed above, the interdependence model has been approached from two points of view. One is the general systems approach that uses individual nation-states as units of analysis.[98] The other approach concentrates on international regimes, such as the law of the sea or international monetary policy,[99] or on individual country case studies.[100] Again, the former focus attempts to discover geographical and/or temporal areas of interdependence under a general theoretical framework defining the concept itself in terms of transactions or their results. The general systems approach relies on empirical analysis to test various theories of interdependence. The second type of study takes the interdependent regime as a given and seeks to explain the mechanisms binding countries together in involuntary partnerships. The split in approach is strikingly similar to the differences noted between the two main streams of international community studies, and it is probably not surprising that this is so. The interdependence model is more a theoretical overlay on the older studies than a separate model giving rise to an entirely new body of work in the field. Thus, it is possible to trace the antecedents of an offshoot model, such as the one to be used in this work, through one or both of the two main branches in the literature over the past three decades.[101]

An Eclectic Model of International Community

As the two main branches of community research have concentrated on different analytical concepts, some sort of accommodation ought to be made in new studies to capture whatever evidence of both kinds of international community exists in a given situation. A two-part model of community that is an elaboration of Richard Cooper's decision to examine international interdependence as a structural and institutional phenomenon can be made, including as much as possible of value from both main streams of international community studies and international interdependence studies.

Structural community shall include those voluntary, subgovernmental linkages between populations that lead to international economic interdependence defined as sensitivity to economic changes in the sense meant by Cooper[102] and to supranational identification as well, predicted by the regional integration model.[103] Thus, structural community has two levels: the first is that of "pathways of contagion," the transactions that create channels along which interdependence can occur; the second is the interdependence itself.[104] In this investigation of structural community among the Arab states, the nations themselves will be used as units of analysis and evidence of high levels of transactions and general economic sensitivity will be sought among them. Economic sensitivity alone is insufficient evidence on which to make judgments about the level of structural community present because economic interdependence can be a negative or a positive relationship.[105] Structural economic interdependence can be individualistic, opportunistic, selfish, and motivated only by a desire for profit. These negative aspects may be ameliorated by transnational identification, such as shared religious or ideological beliefs and individual instances of altruism, gifts of value, or relationships of mutual regard. These create transnational identification on a personal, affective basis. Structural community results from international exchanges of goods and esteem among individuals. These transactions lead to identification among citizens of different countries and channels along which the results of events in one nation can affect another to the extent of influencing domestic politics in the affected nation.[106]

Institutional community is expressed through the formation of international regimes such as common market agreements, international financial institutions, or charters of other kinds of international organizations. Each of these, through joint decisionmaking bodies, allocates resources to accomplish tasks designed to achieve common goals. The range of institutional community goes from alliance to identification. An alliance can be an interstate relationship so distant that resources are allocated only potentially and task decisions are taken entirely at the discretion of the individual partners. Identification places the reponsibility for resource allocation and task assignment on the supranational regime. Where identification is total, no independent parallel activity is carried on in any member state. This situation is similar to the amalgamated security community of Karl Deutsch et al.[107] Extreme alliance behavior allows each partner to keep its national autonomy intact. The extreme of identification surrenders to the international regime complete autonomy within the scope of its statutory jurisdiction. Ecological, that is, naturally occurring, institutional communities tend to fall somewhere between these extremes; the extent of community can be assessed by comparing the relative commitment of members to national and supranational control of resource allocation for commonly held goals. The greater the degree of institutional interdependence, the greater the degree of international community within an international organization.

A related aspect of institutional community requires the international regime to be capable of the type of redistributive policymaking that Mytelka has described as essential for long-term stability.[108] This can be assessed in two ways. One is to find charter provisions for redistribution. The other is to examine the policy output of the regime for evidence of redistribution. The latter is obviously more important, for charter provisions are worthless if never implemented. However, some institutional commitment to redistribution would indicate greater potential stability than merely ad hoc arrangements.

Structural community and institutional community are mutually supportive. Either can precede the other in time. An alliance can be formed with nothing more in common than an enemy; international institutions can be devised to deal with an already highly elaborated structure of transnational relationships. But a high degree of institutional community would seem to depend for its survival, if not for its inception, on extensive structural ties. These ties must be strong enough to dilute the power and the desire of domestic political groups, chiefly the military, to force a reassertion of individual autonomy on the part of member states.[109] Similarly, when structural ties do exist and regardless of the presence or absence of a formal organization binding nations together, domestic pressures can force governments to actions that favor the interests of another nation or a transnational movement more than they help the acting governments.[110] Even where institutional community precedes structural community, the path from alliance to identification is constructed of transnational ties built by the international institution.[111] Significant transnational bonds or affinities between citizens of the several member countries forming an international community are necessary conditions for that community.

Summing up, structural international community is the linkage of populations across national boundaries through perceptions of solidarity, together with tangible and intangible exchanges important enough to influence domestic politics in the affected states. Institutional international community refers to formal bonds between governments that consent to work toward stated goals in concert. Beyond a low level of intergovernmental cooperation, institutional community has a large structural component owing to an increase in interdependence resulting from the international division of labor directed by the international regime. Dependency, or asymmetric interdependence, can occur in an ecological community. One function of the intergovernmental institution(s) is to compensate for dependency through the redistribution of benefits. Failure to deal with dependency can result in the rupture of community bonds.

The role of OAPEC in Arab community will be determined by seeking answers to a set of questions. Is OAPEC a sufficient basis for Arab community? Does it increase the likelihood of cooperative international relations in the area? Is it generally an integrative force or a divisive one in the region—does it complement or compete with other Arab international organizations, principally the Arab League? Finally, what model can this novel organization offer to other international bodies concerned with the regulation of commodity sales and the general level of economic development in their member countries?

Petroleum Cartel

The other aspect of OAPEC to be considered is its economic role in oil-pricing policy. Although the organization formally eschews any active role in setting oil prices (according to its charter, it follows OPEC in all matters regarding international oil policy—see Chapters 2 and 3), a blanket acceptance of its formal position would be naive. The governing body of OAPEC, its Council of Ministers, consists of the identical individuals who represent their countries in OPEC's governing body, the Conference. The oil ministers of each oil-exporting state are important policymakers both domestically and internationally. The fact that they "wear different hats" at different times should not obscure their pivotal positions in oil policy within OAPEC, OPEC, and their national governments. The Arab oil minister who votes on Tuesday in OAPEC to embargo the United States is the same man who votes Wednesday in OPEC to increase the posted price of oil. Thus, an assumption that there might be a concerted influence on OPEC policy exerted by the seven ministers who also sit on the OAPEC Council is plausible. In order to discover this, two points must be clarified. Does OPEC control oil-pricing policy in the international market, that is, is OPEC an oil cartel? And, second, do its Arab members control OPEC? These questions will be answered in Chapter 5. Before that, the cartel concept requires some clarification.

Citizens and policymakers in oil-importing countries have been referring to OPEC as a cartel at least since the autumn of 1973, when oil prices began their most dramatic rise. Members of OPEC regard the cartel appellation as pejorative and incorrect. The purpose of this section is to synthesize a definition of cartel from the economic literature and to demonstrate, in a way intended to be analytical and dispassionate, that OPEC is a cartel in the sense in which the term is used in the United States regarding other industries.

As we shall see, the term "cartel" has carried with it a pejorative connotation since the latter part of the nineteenth century, when it came into general use to describe international "trusts," or agreements among producers of goods in international trade designed to support prices. One might say that there are (at least) two schools of thought about the nature of cartels. One could be called the "strict constructionist" school. Partisans of this viewpoint regard "cartel" as a term that refers to one or more specific forms of market control by producers and that always includes a specific agreement to divide profits and to regulate production in the cartelized industry. This is definitely the minority position among Western scholars, but it is the majority position among the scholars and policymakers of oil-exporting countries. More commonly, "cartel" refers to any association of producers with the object of maintaining or increasing prices. This "loose constructionist" view has dominated scholarship in the West.

International cartels were the subject of extensive research by the League of Nations. According to a 1946 memorandum by league staff member Gertrude Lovasy, such cartels are "voluntary agreements among independent enterprises in a single industry or closely related industries with the purpose of exercising monopolistic control of the market...[C] ommodity agreements and control schemes set up by governments... [are] analogous...."[112] The memorandum does not prescribe a single means or combination of methods of market control as the only way to cartelize an industry.[113] In a general way, the memorandum describes the action of cartels as prevention of competition in markets. "Joint action to prevent competition in certain markets is the main characteristic of international cartels.[114] The raw materials cartels used as cases in the preparation of the memorandum employed a variety of means of keeping prices up: price fixing, production restrictions, buffer stocks, the division of markets, and sales or output quotas. Some cartels provided for penalties to be assessed on violators, while one had only a general agreement to cooperate to increase prices as the basis for its actions for part of its existence as a cartel (the potash cartel). Table 1 lists the cartelized industries Lovasy studied and the kinds of tactics used by each to exert market power. Note that the international petroleum industry is included as one of the cartels. In the interwar period covered by the memorandum, the international oil industry used production restrictions, export quotas, and the division of the international oil market as means of supporting oil prices.[115]

Scholars dealing with intranational restraints of trade have used similarly broad definitions of cartels. "In this country [the United States], [cartel] commonly refers to international marketing arrangements...we have defined cartel as an arrangement among, or on behalf of, producers engaged in the same line of business designed to limit or eliminate competition among them.[116] George Stocking and Myron Watkins modified historical American understanding of the term "cartel" as an international arrangement to include purely domestic agreements to limit competition. The inclusiveness of this definition parallels Lovasy's usage of the term. Stocking and Watkins explicitly include as cartels "loosely defined gentlemen's agreements or informal understandings among business rivals [as well as] formal compacts providing administrative machinery for regulating output, sharing markets, and fixing prices.... National governments occasionally have established them through ... international treaties...."[117]

Ervin Hexner treats directly the difficulty in arriving at an internationally agreed-upon definition of the term "cartel." Because the term was invented (in 1879) by a German to refer to "private market-control mechanisms of

TABLE 1 *Mechanisms of Cartel Action Used by International Raw Materials Cartels Between the First and Second World Wars*

Price Fixing	Buffer Stocks	Sales Quotas	Export Quotas	Market Sharing	Output Restrictions
copper smelting	tin ore	aluminum mercury	cement*	aluminum	copper smelting
aluminum			sulfur	sulfur	lead
sulfur			potash	petroleum	zinc*
			phosphate rock		tin ore
			petroleum		aluminum
			wood pulp		petroleum
			rubber		wood pulp
					rubber

Source: Lovasy, *International Cartels* (Lake Success, 1947), Table 1.

*Formal agreement fixing penalties for failure to abide by cartel policy.

entrepreneurs," the term carried political overtones to French, British, and American scholars.[118] It has often included a pejorative connotation. "[L]abeling a collective market control 'cartel' has almost invariably been to present it in an unfavorable light."[119] For example, the Webb-Pomerene Act, which exempts certain producer associations from the Sherman Antitrust Act, carefully avoids calling such associations cartels, although they certainly conform to the definitions of cartel offered by scholars in the field. The Webb-Pomerene definition has been used explicitly by a former official of OAPEC to deny that OPEC is a cartel and is instead merely a producer association as defined by Webb-Pomerene.[120] The quarrel over whether OPEC is a cartel appears to be tied to the pejorative nature of the term and not to any denial of the fact that OPEC "administers" oil prices.[121]

In a more recent work on industrial organization which has become a standard reference in the field, F. M. Scherer never even defines the term "cartel" in a formal way but rather uses it interchangeably with the term "oligopoly" in situations where the "few sellers" are engaged in collusive practices.[122] Heinrich Kronstein, an international lawyer, in his posthumously published work on the law of international cartels, refers to them as "extensive international private economic orders whose different facets regulate relations between enterprises themselves and . . . relations with third parties.[123] Thus the argument over whether OPEC is a cartel would appear to be specious were it not for the sometimes vociferous protests by citizens and officials of oil-exporting countries at the term and the occasional misunderstanding of the conventional usage of the term by economists and lawyers on the part of scholars in other fields.[124]

Among the several misunderstandings about the term "cartel" when used in reference to OPEC are the following. First, because OPEC does not control all of the oil in the international petroleum market, it cannot be a cartel. Actually, in order for a cartel to be effective in raising prices, all it must control is enough production over marginal levels to maintain its income requirements. In other words, if a cartel must cut total production in order to have an effect on prices, it must control sufficient production to be able to satisfy its members' income needs at a level of production low enough to reduce the total supply of its commodity in the world market to the point that higher prices can be charged.[125] OPEC's cartelization of the oil market was made even easier because demand for oil is inelastic over a wide price range and because a long time is necessary to obtain additional supplies from noncartel sources.[126] This enabled OPEC producers to raise prices very high and to increase their total incomes from oil at lower levels of production.

Another misunderstanding is that the product or commodity controlled by a cartel must be uniform. Because crude oil quality and transportation costs differ greatly from member to member within OPEC and even among different oil fields in the same country, OPEC does not have a uniform product and therefore cannot be a cartel.[127] However, a cartel's need for a uniform product "is obviously true only of quota agreements; there is no cogent reason why mutual market reservation . . . should not be agreed upon, even if the commodities . . . show some variation. Such an agreement simply implies that consumers . . . will have to accept the . . . varieties offered to them."[128] In addition, crude oil is valued not for its intrinsic qualities but for what is made out of it and how those products are used. Oil consumers are buying heating and transportation; they have had to adjust to variations in supplies by lowering air pollution standards for sulfur emissions and by buying lower octane gasoline.

A third misconception is that a cartel must include a formal agreement for allocating production or profits. All the definitions of cartel used here omit such a restriction. Scholars agree that a mechanism for allocating a cartel's benefits adds long-run stability.[129] However, it is not necessary to have such an agreement to call producer collusion a cartel. Historically, cartels are usually organized because of overproduction in an industry, which leads to lower prices through excessive supply and resultant price competition.[130] Thus, one purpose of this kind of cartel is to restrict supply, and cartel agreements reflect this concern. Where excess production is not the main problem leading to cartelization or where producer income can be maintained at a high level despite lower levels of production, formal production agreements are less important for cartel stability. This has been the case for OPEC since 1973.[131]

Finally, there appears to be a real argument over whether governments can form a cartel. According to Hexner, "[T]he term cartel signifies a relationship between entrepreneurs. . . . A market mechanism ceases to be a cartel when it operates in its own right or upon instructions of its government, independent of the volition of its members."[132] But opposed to this definition is

Lovasy's observation: "International cartelization . . . demands that the number of producers in each country be small or that the national industry be organized."[133] Lovasy treated raw materials cartels in which governments were participants identically to those made up solely of entrepreneurs.[134] And, since the success of OPEC in raising oil prices, analysts dealing with other international associations of raw materials producers have used the term "cartel" to describe these organizations.[135] The key to understanding the nature of the individual cartel member is to be able to view each as an independent actor.[136] The member governments of OPEC can be regarded as its independent actors, analogous to the entrepreneurs behind the activities of individual firms. In a later chapter the behavioral independence of the OPEC member governments will be demonstrated in some detail.

One can conclude that it is possible to define OPEC as a cartel because of the following characteristics: it is an organization of producers that has as an object the elimination or amelioration of oil producer competition as it affects oil prices. In addition, the OPEC organization acts as a collusive mechanism for price setting in the international oil market. Simply because a rigid price structure or an agreement on production allocation is missing from the relationship among OPEC members does not disqualify OPEC as a cartel in the sense generally used by economists and international lawyers over the past forty or so years. The key characteristic of individual actor independence, emphasized by Hexner as a necessary condition for a cartel, is fulfilled by the individual government members of OPEC. We shall see that OPEC is not a supranatural organization with coercive power. Therefore it does not, cannot, compromise the independence of its members as economic actors that collude to set the price of oil in the international market.

Notes

1. Important exceptions to this are two works by Zuhayr Mikdashi and one by Karen Mingst. These are: Zuhayr Mikdashi, *The Community of Oil Exporting Countries* (Ithaca, 1972); Mikdashi, "Cooperation Among Oil Exporting Countries with a Special Reference to Arab Countries: A Political Economy Analysis," *International Organization* 28 (Winter 1974); Karen Mingst, "Regional Sectorial Economic Integration: The Case of OAPEC," *Journal of Common Market Studies* 16 (December 1977). In addition to these is Volume 5 (Autumn 1979) of *The Journal of Energy and Development* devoted to OPEC, which contains several articles dealing peripherally with OAPEC, and recent volumes from Middle Eastern authors such as Ali Khalifa Al-Kuwari's book *Oil Revenues in the Gulf Emirates* (Boulder, Colo., 1978). More common is the treatment by Richard C. Weisberg, *The Politics of Crude Oil Pricing in the Middle East, 1970-1975*, Research Series No. 31 (Berkeley, 1977), which delineates OAPEC's role in the oil embargo but does not consider this organization's other roles in the formation of oil policy. Indeed, in a whole volume devoted to the "oil crisis," OAPEC per se is mentioned only briefly in the article contributed by Mikdashi entitled "The OPEC Process." The article by George Lenczowski on "The Oil-Producing

Countries" discussed the embargo without mentioning OAPEC at all. See Raymond Vernon, ed., *The Oil Crisis* (New York, 1976). OAPEC is not even indexed in this volume.

2. *The Middle East Economic Survey (MEES)*, formerly published in Beirut, Lebanon, now in Nicosia, Cyprus, concentrates on economic developments in the entire Middle East region. This weekly publication routinely includes interviews with regional business and political leaders; items culled from the Arab press; the complete texts of documents such as press communiques, company-country contracts; constitutions and resolutions of international organizations in the area; and regular official reports on oil production, oil reserves, oil exports, oil income, and economic development plans for individual countries in the region. *MEES* is invaluable to anyone studying Middle East economic development and is particularly helpful for foreigners who cannot make full use of Arab-language publications. Other important sources are *The Middle East Economic Digest, The Middle East Journal, The Journal of Energy and Development,* and OAPEC publications such as the monthly *Bulletin,* and the annual *Reports of the Secretary General* and *Statistical Reports.*

3. For example, George Stocking's monumental work on Middle Eastern oil, *Middle East Oil: A Study in Political and Economic Controversy* (London, 1970), mentions OAPEC only briefly.

4. This careless kind of analysis was generally confined to newspaper and television reporting. The most interesting "scholarly" work, actually two publications, was by Edward Friedland, Paul Seabury, and Aaron Wildavsky. Their bitter, sustained, violently anti-Arab analysis was first published as a book, but the introductory chapter, entitled "Oil and the Decline of Western Power" can be found in *Political Science Quarterly* 90 (Fall 1975). These two works are the only so-called serious publications that take this conspiratorial viewpoint. M. A. Adelman, a noted petroleum economist, is an example of one who goes to the other extreme and discounts OAPEC's role entirely in the price increases. Adelman blames the United States as much as the oil exporters, Arab and non-Arab alike. (See his "Is the Oil Shortage Real?: Oil Companies as OPEC Tax Collectors," *Foreign Policy* 9 [Winter 1972-73] and Chapter 5 in this volume.)

5. OAPEC includes three Arab oil-exporting countries, Egypt, Bahrain, and Syria, whose oil exports are not high enough to qualify them for membership in OPEC, as well as the seven Arab OPEC members: Saudi Arabia, Kuwait, Iraq, Algeria, the United Arab Emirates (UAE), Libya, and Qatar.

6. In a letter to me dated 2 November 1978, Senator Charles Mathias (R., Md.) wrote "...The question of whether or not Arab members of OPEC have control over cartel policy is very interesting. [The] information on this issue will be very useful...." Senator Mathias is a member of the Committee on Foreign Relations.

7. *The Wall Street Journal,* 16 February 1979, p. 16.

8. Amin Saikal, in his new book *The Rise and Fall of the Shah,* (Princeton, 1980), says that the Shah's position was supported secretly by President Nixon and the then Secretary of State Henry Kissinger as part of the "Nixon Doctrine." This will be discussed in Chapter 5.

9. Mingst, "Regional Sectorial Economic Integration."

10. These countries are also the chief contributors to Third World international aid organizations and development banks.

11. Robert MacDonald, *The League of Arab States* (Princeton, 1965), Chaps. 1-3; M. F. Anabtawi, *Arab Unity in Terms of Law* (The Hague, 1963); or the OAPEC Agreement (Kuwait, 1968).

12. The production cutback was agreed to by the Arab governments the day after the Arabian gulf producers—including Iran—agreed to increase the price of their oil. See J. E. Hartshorn, *Objectives of the Petroleum Exporting Countries* (Nicosia, 1978), p. 9.

13. Walter J. Levy, "The Years That the Locust Hath Eaten: Oil Policy and OPEC Development Prospects," *Foreign Affairs 57* (Winter 1978/79).

14. Ibid.; Al-Kuwari, Oil Revenues in the Gulf Emirates, Chaps. 8-9; Ibrahim M. Oweiss, "Strategies for Arab Economic Development," *Journal of Energy and Development 3* (Autumn 1977); and countless public speeches by oil and development ministers from the OPEC and OAPEC countries.

15. Ibid.; Peter Beaumont, Gerald Blake, and Malcolm Wagstaff, *The Middle East* (New York, 1976).

16. David Landes, *Bankers and Pashas* (New York, 1958); Robert Mabro and Samir Radwan, *The Industrialization of Egypt, 1939-1973* (Oxford, 1976).

17. Ibid.

18. Mary Ann Tétreault, "Foreign Aid and the Egyptian-Israeli Peace Treaty: The Case of Egypt" (Paper presented at the International Studies Association annual meeting in Los Angeles, 19-22 March 1980), p. 2.

19. Ragaei El Mallakh, "Industrialization in the Arab World: Obstacles and Prospects," in Naiem Sherbiny and Mark Tessler, eds., *Arab Oil* (New York, 1976), p. 58; also Levy, "The Years That the Locust Hath Eaten," pp. 296-99.

20. These countries are Saudi Arabia, Libya, Kuwait, Qatar, and the UAE. The financial inflows were quickly spent by most of these nations, and a gap gradually appeared as the real price of oil declined prior to price increases in 1979.

21. Beaumont et al., *The Middle East*, p. 179; and the most recent census, reported in MEES, 12 May 1980, which listed the population of Kuwait as 562,065 nationals and 793,762 foreigners.

22. Interview with Bader O. Aldafa, first secretary, Qatar Embassy, Washington, D.C., 23 May 1980.

23. Ali Jaidah, "Downstream Operations and the Development of OPEC Member Countries," *Journal of Energy and Development 4* (Spring 1979) 304-6.

24. A. B. Zahlan, *Science and Science Policy in the Arab World* (New York, 1980), p. 57.

25. Service industries are the logical subsequent step in development because of the paucity of resources other than petroleum in most Arab oil-exporting countries. (See Beaumont et al., The Middle East, pp. 234-35.)

26. Zahlan, *Science and Science Policy,* p. 130.

27. Beaumont et al., *The Middle East.* Chap. 7; also Ali A. Attiga, "Regional Cooperation in Downstream Investments: The Case of OAPEC" (Paper delivered at the OPEC Seminar on the Present and Future Role of the National Oil Companies, Vienna, 10-12 October 1977. Dr. Attiga is the secretary general of OAPEC and his paper criticized the lack of such coordination. Walter Levy made the same point in "The Years That the Locust Hath Eaten."

28. Zahlan, *Science and Science Policy,* Chap. 1.

29. "In advanced countries . . . the activity is obviously internalized . . . [rather than] externalized . . . " (ibid., p. 19).

30. Jaidah, "Downstream Operations," p. 306.

31. Levy, "The Years That the Locust Hath Eaten," p. 299. Levy lists reasons ranging from transportation costs and economic bottlenecks to the necessity to pay bribes to win a contract. OAPEC country purchasers regard the higher costs as a function of corporate greed and developed country desires to recoup foreign payments made for oil imports. See recent editorials in the monthly OAPEC publication *Bulletin*, June 1979, February 1980, and March 1980. Editorials are printed on the front cover of each issue.

32. Levy, "The Years That the Locust Hath Eaten," pp. 296-99.

33. This difference is emphasized by Malcolm Kerr, "Regional Arab Politics and the Conflict with Israel," in Paul Hammond and Sidney Alexander, eds., *Political Dynamics in the Middle East* (New York, 1972).

34. This is the thesis of Arnold Hottinger, *The Arabs: Their History, Culture and Place in the Modern World* (Berkeley, 1963), and of Joel Carmichael, *The Shaping of the Arabs* (New York, 1967).

35. Hottinger offers this explanation, which is supported in works by Arab scholars, such as Albert Hourani, "Race, Religion, and Nation State in the Near East," in *A Vision of History* (Beirut, 1961); George Antonius, *The Arab Awakening* (London, 1938); and American political scientists such as Michael C. Hudson, *Arab Politics* (New Haven, 1977), Chap. 2.

36. Landes, *Bankers and Pashas;* and Hottinger, *The Arabs,* Pts. 2 and 3.

37. Antonius, *The Arab Awakening*, Chaps. 1-3; Anabtawi, *Arab Unity in Terms of Law,* Chaps. 1-2.

38. This nation would have been located in the Fertile Crescent and would have omitted Egypt as well as most of the Arabian Peninsula outside the Hejaz. See the documents in the appendix to Antonius, *The Arab Awakening.*

39. See Antonius, *The Arab Awakening,* Anabtawi, *Arab Unity in Terms of Law,* and Hottinger, *The Arabs.* Antonius reprints all the documents concerning promises made to the Arabs and what was available in 1938 concerning the other agreements of the period. Antonius, a young Christian scholar, was privileged to have been able to talk to Sharif Husain and his sons and to have access to their contemporary political papers in the preparation of his excellent work.

40. Kerr, "Regional Arab Politics and the Conflict with Israel," p. 34.

41. Hottinger, *The Arabs,* p. 222.

42. The effects of French economic policy have been long-lasting. Trade regulations and trading patterns in the former French colonies still reflect a dependence on the metropolitan economy.

43. Amal Vinogradov, "The 1920 Revolt in Iraq Reconsidered: The Role of Tribes in National Politics," *International Journal of Middle East Studies* (April 1972). Vinogradov argues that state-centered nationalism as well as Arab nationalism existed together in Iraq from the beginning of the British occupation. Hottinger did not discuss state-centered nationalism in his treatment of the 1920 revolt.

44. Hottinger, *The Arabs,* p. 233.

45. P. J. Vatikiotis, "The Politics of the Fertile Crescent," in Hammond and Alexander, eds., *Political Dynamics in the Middle East,* p. 251.

46. The Hejaz, a narrow strip of territory along the western coast of the Arabian Peninsula, was under Turkish rule during the colonial period, but the vast area of the Arabian Peninsula was left alone.

47. Hottinger, *The Arabs*, Pt. 2. The Wahhabis were followers of Muhammad ibn Wahhab, an eighteenth-century Islamic teacher who preached that Islam had become corrupt and required purification. Wahhab joined with Sa'ud of Dav'iya to extend a political-religious hegemony over the tribes of the Arabian Peninsula. The Wahhabi kingdom was attacked and destroyed by Muhammad Ali of Egypt in 1818.

48. Ibid.

49. Landes, *Bankers and Pashas,* Chap. 3. Even before the introduction of long staple cotton as a cash crop in the early nineteenth century, Egyptian peasants, the fellahin, were a rich source of tax money.

50. Kerr, "Regional Arab Politics," p. 34.

51. The Arab Nationalist Movement, an organized transnational group whose aim was to form an Arab nation in the Fertile Crescent that would have included Palestine—not Israel—was most powerful in the Fertile Crescent area. The Arab Nationalist Movement was a threat to the legitimacy of the Fertile Crescent states. It eventually broke into factions, each of which became more closely identified with the grievances of the Palestinians. Several Arab Nationalist Movement leaders were Palestinians. See W. W. Kazziha, *Revolutionary Transformation in the Arab World* (New York, 1975).

52. Kerr, "Regional Arab Politics"; Malcolm H. Kerr, "The United Arab Republic: The Domestic, Political and Economic Background of Foreign Policy," in Hammond and Alexander, eds., *Political Dynamics in the Middle East.*

53. Anabtawi, *Arab Unity in Terms of Law;* MacDonald, *The League of Arab States.*

54. Michael C. Hudson regards the issue of the shaky legitimacy of the individual Arab governments as critical to an understanding of the politics of the entire region. *Arab Politics,* Chap. 1.

55. Kerr, "Regional Arab Politics."

56. Donald L. Losman, "The Arab Boycott of Israel," *International Journal of Middle East Studies* 3 (April 1972).

57. This is the publicly expressed opinion of Western scholars such as Kerr and MacDonald. Arab scholars such as Anabtawi violently disagree. Kerr believes that mass support for Arab unity is the major, and in some cases the only, factor that underlies statements of Arab officials supporting unity. MacDonald believes that Arab unity as other than an emotional, mass-level issue has been dead since World War II.

58. A serious attempt is one that gets beyond the "let's do it" stage to actual implementation. The United Arab Republic did have a unified government under Nasser. But economic difficulties were probably at the root of its failure in the eyes of the Syrians. See E. Kanovsky, "Arab Economic Unity," in Joseph Nye, ed., *International Regionalism* (Boston: 1968), p. 375.

59. Fayez A. Sayegh, *Arab Unity: Hope and Fulfillment* (New York, 1958), is a good example of Arab hopes during the partial unification period.

60. Limited partial unification is unification of less than the entire polity of less than the total number of Arab states, such as a customs union of less than all the members of the Arab League. OAPEC is an example of limited partial unification.

61. The confrontation states, located on the border of Israel, are Egypt, Syria, and Jordan. There is reason to believe that King Faisal of Saudi Arabia knew and approved of Egypt's plans because of the unusual and frequent conferences between Faisal and Egyptian President Sadat in September and because of Faisal's subsequent support of the embargo and cutback policy begun on 17 October. (See the weekly issues of *MEES* for reports of the conferences and the genesis of the OAPEC policies of October 1973.)

62. Christopher T. Rand, *Making Democracy Safe for Oil* (Boston, 1975), Chap. 16.

63. Kerr, "Regional Arab Politics."

64. The evidence for this is anecdotal but fairly widespread. For example, Thomas Kiernan, *The Arabs: Their History, Aims and Challenge to the Industrialized World* (Boston, 1975), Preface; Jules Archer, *Legacy of the Desert* (Boston, 1976). From my own anecdotal experiences, I have come to believe this is not the way Saudis see themselves but that it is common for nationals of other Arab countries.

65. Bruce M. Russett, *International Regions and the International System* (Chicago, 1967).

66. Karl W. Deutsch, Sidney A. Burrell, Robert A. Kann, Maurice Lee, Jr., Martin Lichterman, Raymond E. Lindgren, Francis L. Loewenheim, and Richard W. Van Wagenen, *Political Community and the North Atlantic Area* (Princeton, 1957).

67. Karl Kaiser, "The Interaction of Regional Subsystems," *World Politics* 21 (October 1968).

68. Russett, *International Regions*, pp. 167-90.

69. Charles W. Kegley and L. Howell, "The Dimensionality of Regional Integration: Construct Validation in the South East Asia Context," *International Organization* 29 (Autumn 1975). This work was intended to be a test of Joseph Nye's typology of integration sectors (*Peace in Parts* [New York, 1971]), but I believe the operationalization of the variables in this case was not adequate to serve as a proper test of the Nye typology.

70. Bruce M. Russett, "International Transactions, Community and Integration," *Journal of Common Market Studies* 9 (March 1971); and P. Savage and K. Deutsch, "A Statistical Model of the Gross Analysis of Transaction Flows," *Econometrica* 28 (1960).

71. Roger Cobb and Charles Elder, *International Community: A Regional and Global Study* (New York, 1970).

72. Ernst Haas, *The Uniting of Europe* (Stanford, 1958); E. Haas and P. Schmitter, "Economics and Differential Patterns of Political Integration," *International Political Communication* (Garden City, N.Y., 1966); Phillippe Schmitter, "A Revised Theory of International Integration," *International Organization* 24 (Autumn 1970).

73. For example, Donald Puchala, "International Transactions and Regional Integration," *International Organization* 24 (Autumn 1970).

74. Ronald Inglehart, "Public Opinion and Regional Integration," *International Organization* 24 (Autumn 1970).

75. Schmitter, "A Revised Theory of International Integration." This mechanism is called "spillover."

76. Ibid. This is "spillback."

77. E. Haas, "The Study of Regional Integration: Reflections on the Joy and Anguish of Pre-Theorizing," *International Organization* 24 (Autumn 1970).

78. Ibid.; also Leon Lindberg and Stuart Scheingold, *Europe's Would-Be Polity* (Englewood Cliffs: 1970).

79. Donald J. Puchala, "Domestic Politics and Regional Harmonization in the European Communities," *World Politics* 27 (July 1975).

80. Louis J. Cantori and Stephen L. Spiegal, "The Analysis of Regional International Politics: The Integration Versus the Empirical Systems Approach," *International Organization* 27 (Autumn 1973).

81. Roger D. Hansen, "Regional Integration: Reflections on a Decade of Theoretical Efforts," *World Politics* 21 (January 1969). Hansen took the more traditional view that international relations could be divided plausibly into "high politics" regarding peace and war, and "low politics" dealing with economic matters.

82. Haas, "Reflections on the Joy and Anguish"; Schmitter, "A Revised Theory"; Ernst B. Haas, "Turbulent Fields and the Theory of Regional Integration," *International Organization* 30 (Summer 1976); and Joseph Nye, "Comparing Common Markets," *International Organization* 24 (Autumn 1970). Nye and Robert Keohane were already investigating nongovernmental interstate relations by looking at the roles of nonstate actors in international relations (see Nye and Keohane, eds., *Transnational Relationships and World Politics* [Cambridge, Mass., 1972]).

83. Richard N. Cooper, *The Economics of Interdependence and Economic Policy in the Atlantic Community* (New York, 1968); and Richard N. Cooper, "Economic Interdependence and Foreign Policy in the 1970's,"*World Politics* 24 (January 1972).

84. Cooper, *The Economics of Interdependence,* Chap. 1, and "Economic Interdependence and Foreign Policy."

85. Ibid.

86. For example, by Robert Keohane and Joseph Nye, *Power and Interdependence* (Boston, 1977), pp. 11-19; Richard Rosecrance, et al., "Whither Interdependence?" *International Organization* 31 (Summer 1977), pp. 426-29.

87. Cooper, "Economic Interdependence and Foreign Policy." His viewpoint is shared by Edward Morse, "Crisis Diplomacy, Interdependence and the Politics of International Economic Relations," *World Politics* 24 (Spring 1972).

88. Haas, "Turbulent Fields."

89. Aaron Segal, "The Integration of Developing Countries: Some Thoughts on Africa and Central America," *Journal of Common Market Studies* 2 (June 1967); Lynn K. Mytelka, "Fiscal Politics and Regional Redistribution, " *Journal of Conflict Resolution* 19 (March 1975). Mytelka's theories are set out at length in an article by her former teacher, William Axline, "Underdevelopment, Dependency and Integration: The Politics of Regionalism in the Third World," *International Organization* 31 (Winter 1977).

90. Such as William Loehr, "Collective Goods and International Cooperation," *International Organization* 27 (Summer 1973); or Stephen Brown, David Price, and Satish Raichur, "Public Good Theory and Bargaining Between Large and Small Countries," *International Studies Quarterly* 20 (September 1976).

91. Richard Musgrave and Peggy Musgrave, *Public Finance in Theory and Practice* (New York, 1973), Chap. 3; and M. Peston, *Public Goods and the Public Sector* (London, 1972).

92. Mancur Olson, *The Logic of Collective Action* (Cambridge, Mass., 1965); and Mancur Olson and Richard Zeckhauser, "An Economic Theory of Alliances," *The Review of Economics and Statistics* 3 (August 1966).

93. Theodore J. Lowi, "American Business, Public Policy, Case Studies and Political Power," *World Politics* 16 (July 1964); *The End of Liberalism* (New York, 1969); and "Four Systems of Policy, Politics and Choice," *Public Administration Review* 32 (July/August 1972).

94. Mytelka as described in Axline, "Underdevelopment, Dependency and Integration."

95. Kanovsky, "Arab Economic Unity."

96. Mytelka as described in Axline, "Underdevelopment, Dependency and Integration," Model 3.

97. Lynn Mytelka, "Fiscal Politics and Regional Redistribution."

98. Roger Tollison and Thomas Willett, "International Integration and the Interdependence of Economic Variables," *International Organization* 27 (Spring 1973); Peter J. Katzenstein, "International Interdependence: Some Long-term Trends and Recent Changes," *International Organization* 29 (Autumn 1975); and Rosecrance et al., "Whither Independence?"

99. Keohane and Nye, *Power and Interdependence.*

100. Gary Gereffi, "Drug Firms and Dependency in Mexico: The Case of the Steroid Hormone Industry," *International Organization* 32 (Winter 1978). Dependency is asymmetric interdependence. See Mary Ann Tétreault, "Measuring Interdependence," *International Organization* 34 (Summer 1980).

101. A similar analysis of what he calls the "global relations literature" was done by James A. Caporaso, "Dependence and Dependency in the Global System," *International Organization* 32 (Winter 1978). Caporaso finds four types of global relations literature according to the level of analysis used in each study (see pp. 34-42).

102. Cooper, "Economic Interdependence and Foreign Policy."

103. See, for example, Donald J. Puchala, "International Transactions and Regional Interdependence."

104. This distinction is similar to that of Rosecrance et al., who define "horizontal interdependence" as transactions and "vertical interdependence" as the response of one economy to changes in another.

105. Haas, "Turbulent Fields."

106. This domestic outcome of interdependence is discussed by Cooper, "Economic Interdependence and Foreign Policy"; Keohane and Nye, *Power and Interdependence*; and Gereffi, "Drug Firms and Dependency."

107. *Political Community*, 5-7, although it is confined only to the sector supervised by the international regime.

108. Mytelka as described in Axline, "Underdevelopment, Dependency and Integration," Models 2 and 3.

109. This is precisely what happened in the case of Egyptian-Syrian union to form the United Arab Republic. See Kanovsky, "Arab Economic Unity."

110. Gereffi, "Drug Firms and Dependency." This is also implied by Malcolm Kerr, "Regional Arab Politics."

111. This is the spillover argument referred to previously (Schmitter).

112. Gertrude Lovasy, *International Cartels*, Department of Economic Affairs, League of Nations (Lake Success, N.Y.: 1947), p. 1. Lovasy is credited with authorship in the Introduction. An earlier League of Nations study by Paul de Rousiers, *Cartel and Trusts, Their Development*, League of Nations Economic and Financial

Section (Geneva, 1927), calls a cartel a voluntary agreement made freely between manufacturers for a specific end (pp. 1-2). This is even more general than Lovasy's definition.

113. Lovasy, *International Cartels*, p. 2.

114. Ibid., p. 1.

115. For a detailed treatment of the tacit collusion and overt cartel arrangements of the international petroleum industry that includes this interwar period, see *The International Petroleum Cartel*, Staff Report by the Federal Trade Commission submitted to the Senate Subcommittee on Monopoly of the Select Committee on Small Business (Washington, D.C., 1952), pp. 191-348.

116. George W. Stocking and Myron W. Watkins, *Cartels in Action* (New York, 1947), p. 1.

117. Ibid.

118. Ervin Hexner, *International Cartels* (London, 1946), p. 1.

119. Ibid., p. 5.

120. The official is George Tomeh, a highly respected scholar in his own right, and he made this point in a letter to me dated 13 April 1978. Of course, Dr. Tomeh is correct. OPEC can be defined as a Webb-Pomerene producers' association, and the fact that Westerners call it a cartel has indeed a pejorative intent. My point, however, is that OPEC, as well as every other Webb-Pomerene producers' association, is also a cartel in the sense meant by perfectly sober economists with no political axe to grind with regard to the oil-exporting countries.

121. I say this because in order to correct my view that OPEC is a cartel Dr. Tomeh had sent to me a paper by Fadhel Al-Chalabi, "Pricing of OPEC Crude Oil: A Case for the Valuation of Depletable Resources in Relation to Economic Development," manuscript (Kuwait, n.d.). Chalabi makes a straightforward analysis of OPEC's function as a price-administering organization while simultaneously taking issue with the cartel designation. This is an excellent example of the "strict constructionist" point of view. Another example is Zuhayr Mikdashi, "The OPEC Process," in Vernon, ed., *The Oil Crisis*, pp. 205-6, and Z. Mikdashi *The International Politics of Natural Resources* (Ithaca, 1976), pp. 78-79. It is possible that the influence of the strict constructionist viewpoint in the West is due to the respect in which Mikdashi is held. A distinction among types of cartels similar to the strict constructionist-loose constructionist dichotomy is made by Richard Caves, "International Cartels and Monopolies in International Trade," Harvard Institute of Economic Research, Discussion Paper Series (Cambridge, Mass., July 1977), pp. 4-5. Caves calls the strict constructionist cartel a "contractually complete cartel" but refers to all cases of collusive behavior among producers as cartels.

122. F. M. Scherer, *Industrial Market Structure and Economic Performance* (Chicago, 1970).

123. Heinrich Kronstein, *The Law of International Cartels* (Ithaca, 1973), p. 4.

124. When an earlier version of Chapter 5 of this work was presented at the Annual Meeting of the Southwest Political Science Association in 1978, a Libyan in the audience approached me afterward with the same objections I had had from Dr. Tomeh. Gabor Galantai, professor of government at the University of Puerto Rico and, in 1977-78, visiting scholar at the Center for Middle Eastern Studies at the University of Texas, Austin, is also a strict constructionist and criticized this chapter

on the same grounds. The strict constructionist viewpoint was articulated very well in a manuscript by Lawrence Schultz, a former student of Dr. Galantai, "Dead Theories Surviving the Patient" mimeographed (n.p., n.d.), which held that OPEC has not collapsed as predicted by some economists (most notably M.A. Adelman in "Is the Oil Shortage Real?") precisely because it is not a cartel. My own view of why the OPEC cartel has not collapsed is discussed in Chapter 5.

125. Precisely this kind of analysis has been made with reference to OPEC by Paul Leo Eckbo, *The Future of World Oil* (Cambridge, Mass., 1975), Chaps. 1, 6.

126. S. D. Krasner, "Oil Is the Exception," *Foreign Policy* 14 (Spring 1974); and Z. Mikdashi, "Collusion Could Work," ibid. Even when additional production becomes available, there is no guarantee that proprietors of the new supply won't charge the cartel price. See Øystein Noreng, "Friends or Fellow Travelers? The Relationship of Non-OPEC Exporters with OPEC," *Journal of Energy and Development* 4 (Spring 1979); and Chapter 5.

127. This is a point made by Lawrence Schultz in the paper cited in note 124.

128. Lovasy, *International Cartels*, pp. 2-3.

129. See Scherer, *Industrial Market Structure*, Chap. 6; R. A. Smith, "The Great Electrical Conspiracy," *Fortune*, May 1961; Eckbo, *The Future of World Oil*, Chap. 6.

130. Stocking and Watkins, *Cartels in Action;* Paul W. MacAvoy, *The Economic Effects of Regulation* (Cambridge, Mass., 1965). MacAvoy's book is a detailed study of the several attempts to cartelize the American railroad industry in post-Civil War America (and how government regulation under the Interstate Commerce Commission, created in 1887, substituted a public economic regime to hold up prices for these formerly private efforts). Lovasy makes this same point in *International Cartels*, pp. 5-9.

131. Zuhayr Mikdashi discusses the several attempts of OPEC to arrive at a production-sharing agreement before 1973. Mikdashi, *The Community of Oil Exporting Countries*, Chap. 5.

132. Hexner, *International Cartels*, pp. 19-20.

133. Lovasy, *International Cartels*, p. 3. Stocking and Watkins share this viewpoint (*Cartels in Action*, p. 4).

134. Lovasy, *International Cartels*, Table 1. Refer back to Table 1 in the text.

135. Caves, *International Cartels and Monopolies*. Also Robert S. Pindyck, "Cartel Pricing and the Structure of the World Bauxite Market," *The Bell Journal of Economics* 8 (Autumn 1977); and by the same author, "Gains to Producers from the Cartelization of Exhaustible Resources," mimeographed (n.p., January 1976); Davis B. Bobrow and Robert T. Kudrle, "Theory, Policy and Resource Cartels: The Case of OPEC," *Journal of Conflict Resolution* 2 (March 1976). While this last work concentrates on OPEC, it makes the point that similar organizations are cartels in the same sense.

136. Hexner, *International Cartels*, pp. 19-20. The quality of *independence* rather than the nature of the individual actor-participant as entrepreneur or as government is the primary reason for Hexner's insistence that cartel members be entrepreneurs. He did not envision governments playing similar roles. Lovasy did (*International Cartels*, p. 1).

Chapter 2

International Organizations and Arab Petroleum Policy

Two main themes describe the course of Arab politics in the twentieth century. The most recent reflects intra-Arab divisions aggravated by the spread of revolutionary regimes among the nations in the region. These divisions were so deep and acrimonious that during the period 1957-67 scholars used the term "Arab Cold War" to describe the situation.[1] Yet preceding, underlying, and surviving periods of major intra-Arab dissention runs the theme of Arab unity. Arab unity traditionally referred to political unification of the Arab states into one Arab nation.[2] It has also served as the theme of international and transnational collective action for other goals.

The Arab League

The League of Arab States, commonly called the Arab League, was formed in March 1945 "to supervise the execution of agreements made by member states, to supervise in a general way the interests of Arab countries, to coordinate their political plans and to insure cooperation among them, and to protect their sovereignty and independence."[3] Thus the aims of the Arab League regarding unity in the sense of political unification were modest. M. F. Anabtawi is critical of the Arab League pact in part because it omitted any statement or provision calling for Arab unification as an aim of the Arab League.[4] But another student of the Arab League has pointed out that even the Alexandria Protocol, a more radical document than the Arab League pact, did not suggest that all-inclusive Arab unification was a feasible political goal.[5] The founders of the Arab League regarded their mandate as a call for the creation of a regional organization to serve as a vehicle for Arab community, not as a call for the formation of a unitary Arab state.[6] A guarantee to respect sovereignty permitted members to feel secure in surrendering some autonomy to collective decisionmaking. Those provisions of the Pact of the Arab League that appear to some to be inimical to interstate unification via the league, are regarded by others as the foundation for amicable cooperation among the members. For example, Article 7 of the league pact states that "majority decisions shall be binding only upon those states which have accepted them," and Articles 2 and 8 leave final decisions on cooperation among member states, even those participating in unanimous decisions, to the individual governments of the states. These provisions have been interpreted both as surrender to the status quo[7] and as guarantees of national independence.[8]

Robert MacDonald presents the Arab League as a community-building organization rather than as a means for promoting regional political integration. It is "the agency through which the Arab states have consciously confirmed their own national sovereignties and secured their region from external control."[9] This is a signal achievement whether Arab sovereignty is distributed across several states or concentrated in only one. The Arab League has also served as a vehicle for coordinating the positions taken by its members in United Nations organs in which they participate. Group voting cohesion within the Arab bloc in the United Nations is well documented.[10] The Arab League also coordinates an anti-Israel boycott which many regard an effective instrument of economic harassment.[11] Regional economic, social, and cultural integration has progressed steadily, if slowly, and "millions of Arabs have had their lives influenced by the activities of the Arab League in such fields as education, science, economics, medicine and public health, and cultural affairs."[12] Functional interdependence among Arab states has been fostered and nourished by the Arab League. That Arab states have not deserted the league, especially in view of the extreme political incompatibility among nations in the region, demonstrates its functional utility to its members.[13]

Despite these achievements, the Arab League has not been able to perform effectively in two kinds of situations: the first is when its members are divided on an issue. The Arab League is very large. Today it has more than twenty members on two continents. They include a variety of political regime types, economic systems, and colonial histories. When only one or two differ from the rest, consensus is possible even on very sensitive issues.[14] But when members divide more evenly, or when the dissenters can turn an issue into one of member sovereignty, collective policymaking is impossible.

The other type of situation in which the league is an inadequate instrument for collective policymaking is when non-Arab nations have to be included for policy to be effective. Oil politics combines both situations. In late 1959 and 1960, when excess supplies of petroleum were exerting downward pressure on oil prices, the need for an effective forum for coordinating member nations' oil policies became acute. The chief problem with using the Arab League for such a purpose lay in the fact that two of the major oil-exporting nations, Iran and Venezuela, were not Arab countries and were thus excluded from decisionmaking in the league. Although it might have been possible for the Arab League to apply political pressure on the oil companies through a collective policy of its own, the existence of significant oil production capacity outside the organization would have limited if not prevented any exercise of market power.

The Arab League had considered using oil as a political weapon against Israel as early as February 1946.[15] It established the Committee of Oil Experts in 1952 to devise means of protecting the political independence and territorial integrity of Arab states through coordination of their oil policies.[16] By 1959, however, the principal issues facing oil exporters in the Middle East and elsewhere had to do with economics: oil prices and the incomes these nations were able to generate through oil exports.

Since its foundation in 1945 the Arab League had considered creating an Arab petroleum organization. However, the member nations realized that in order for an organization with economic goals to be able to bargain effectively with the oil companies "the assoication envisaged had to include the non-Arab large exporters of petroleum, notably Iran and Venezuela."[17]Reflecting Arab appreciation of the economic role of these nations, Iran and Venezuela were invited to attend the First Arab Petroleum Congress, sponsored by the Arab League in 1959, as observers.[18]

The congress served as a forum for the expression of dissatisfaction over the behavior of the oil companies,[19] which, beginning in Feburary 1959, had unilaterally reduced the posted, or tax reference, prices for crude oil by about eighteen cents per barrel.[20] Influenced by lobbying on the part of Perez Alfonzo, minister of mining and hydrocarbons for Venezuela, and at the urging of Abdullah Tariki, Saudi Arabian minister of petroleum and minerals, the formation of an oil consultation commission to include Iran, Venezuela, the Arab exporting countries, and the Arab League itself, was proposed and adopted by the congress. While it existed, the commission advocated the stabilization of posted prices for crude oil exports and recommended that Middle Eastern governments raise their tax take on oil exports, thus reducing, at least in the short run, the cost advantage of Middle Eastern over Venezuelan crude oil.[21] However, the Oil Consultation Commission was shortlived, reportedly owing to intra-Arab quarrels.[22]

In August 1960, the companies cut the posted price of crude oil by another ten cents per barrel, again without consulting the host governments. The seriousness of this action to host country budgets can be seen in the results of sample calculations of revenue loss projections due to the August 1960 cut. The imputed "loss of tax proceeds" to the governments in the Middle East amounted to four cents per barrel or about $300 million for exports from the area during the period August 1960 through the end of 1963.[23] This second price reduction precipitated a crisis meeting in Baghdad of representatives of the governments of Iran, Iraq, Kuwait, Saudi Arabia, and Venezuela in September 1960. The representatives agreed to create a permanent intergovernmental organization of petroleum-exporting countries.

OPEC

The new organization, called the Organization of Petroleum Exporting Countries—OPEC—set out its aims in its first resolution: to maintain prices for crude oil free from all unnecessary fluctuations; to restore oil prices to their pre-February 1959 levels; to induce the oil companies to consult the host governments before any decisions on price modifications should be made; and eventually to develop a program "to ensure stable oil prices by, among other means, the regulation of production."[24] Toward these ends, OPEC would work on the basis of unanimous consent among its members.

OPEC members believed that it was legitimate for them to influence oil prices because they regarded the existing petroleum price structure as the outcome of oil company policy and not as the result of market forces.[25] They hoped, by collective action, to exert a more forceful claim to the economic rents (profits over the "normal" profit level in the classical competitive model) which the oil companies were keeping for themselves.[26] During the 1960s, however, OPEC found it difficult to exert collective pressure directly on the oil companies. The companies refused to treat with OPEC as a legitimate spokesman for the member nations and insisted on negotiating with each nation individually.[27] In 1962 the Conference, OPEC's decisionmaking organ, designated the secretary general of the organization as its official negotiator, empowered to treat with the oil companies on matters such as royalty expensing [28] where a joint policy had been agreed upon in OPEC resolutions. But due to the refusal of the companies to recognize him as the legitimate representative of the host governments, the countries had to deal individually with the companies.[29]

In spite of this, the ability to arrive at a unified policy via the OPEC process was probably a major factor accounting for what success the countries did enjoy during this period. They were able to maintain the level of tax reference prices and also to obtain a variety of financial concessions from the companies.[30] They achieved agreements on the gradual elimination of the right of the companies to deduct royalties from the amount they paid in taxes to the host governments—royalty expensing;[31] a series of agreements designed to eliminate the practice of giving the companies various discounts off the posted prices; agreements maintaining gravity differential allowances to the countries, increasing the take to those countries that produced crudes yielding more profitable products after refining;[32] and finally, agreements reducing the size of the marketing allowances the countries ceded to the companies, ostensibly to cover the costs of marketing their oil.[33] Negotiations on these issues were conducted throughout most of the 1960s. Although no single agreement represented a bonanza for the host countries, each one served to preserve and gradually to increase the income they received per barrel of oil, even though posted prices did not return to their pre-February 1959 levels (see Table 2).

Additionally, the negotiation process itself was a testing ground for OPEC. Its members did not always agree to accept company offers after a round of negotiations. For example, during the negotiations on royalty expensing, the

TABLE 2 Payments to Host Governments in Cents/Barrel of Exports

Year	Country							
	Kuwait	Saudi Arabia	Iran	Iraq	Abu Dhabi	Qatar	Libya
1960	76.5	75.0	80.1	78.6	—	86.4	—
1961	74.4	75.5	75.8	76.5	—	83.0	62.7
1962	74.8	76.5	74.5	76.7	50.9	82.3	64.7
1963	74.3	78.7	79.7	80.7	36.4	84.2	65.1
1964	76.9	82.0	80.9	80.1	18.2	84.4	62.9
1965	78.9	83.2	81.1	81.7	32.5	82.2	83.8
1966	78.4	83.4	81.4	81.3	74.3	87.2	87.0
1967	79.3	84.8	82.5	85.2	76.3	87.2	101.6
1968	80.5	87.8	83.7	90.7	84.5	88.1	100.7
1969	80.8	87.1	80.9	91.4	87.3	91.9	100.0
1970	82.8	88.3	86.2	95.7	92.0	91.5	109.0

Source: Table derived by Marwan Iskander from publications of "The Petroleum Information Foundation, Inc.", as reported in Marwan Iskander, *The Arab Oil Question* (Beirut: 1974), pp. 132-33.

Note: Payments are obligations for royalty and income tax for the year shown, including small amounts arising from export refining operations in Saudi Arabia, Kuwait, and Iran. Retroactive payments are attributed to the year applicable. Payments were reported in pounds sterling and converted at $2.50/pound.

oil companies offered a proposal that would have reduced the increase in host country take by introducing a new series of discounts in exchange for agreeing to royalty expensing. Five OPEC members (Iran, Kuwait, Qatar, Libya, and Saudi Arabia) made it known at the Seventh Conference in November 1964 that they felt the offer was the best obtainable without resorting to coercive acts against the companies. Other members (Iraq, Venezuela, and Indonesia) opposed this position very strongly. They were dissatisfied with the financial return they would get under the proposal and also opposed it on the grounds that it would result in their relinquishing sovereignty to the companies, which were specified in the proposal as sole judges of whether or not to reduce the discounts after three years. However, since Iraq alone was actually a party to the negotiations (which applied only to the Middle Eastern producers) the conference decided to "recognize the freedom of action of each member country in regard to the issue concerned."[34]

This position was significant. First, it represented a way for OPEC to avoid internal conflict when a small faction within its membership could not be reconciled to a consensus of the rest: each member should be permitted to act according to its own preferences and still be able to remain within the organizational fold. Instead of a comprehensive, OPEC-wide decision on divisive issues, the organization could go as far as its institutional requirement for unanimity would permit and beyond that allow members to pursue individually variant courses. Second, this policy had significance externally, or outside OPEC, becuase it virtually institutionalized the practice of leap-frogging. "Leapfrogging" is the name given to the negotiating tactic by which one state exacts as much as it can get by whatever means it has; upon its success at obtaining more favorable terms than those enjoyed by its fellows, the fellows demand equal treatment. OPEC fostered such behavior by permitting members to *exceed* its minimum policy standards. The companies themselves had already laid the groundwork for this permissive policy by requiring each host country to negotiate individually with companies operating on its territory. Any gain made by one nation could then become the floor demand for subsequent negotiations. This technique was used most effectively by Libya beginning in 1970 to achieve price increases of unprecedented magnitude.[35]

Up until the June 1967 Arab-Israeli war, the accomplishments of OPEC on the price front, though limited, were of some importance. The exporting governments were able to increase their take per barrel of oil in a period of world oil glut. OPEC itself became stronger with every member victory over the oil companies. Even though each member had to negotiate individually with its concessionaires, it negotiated on the basis of an OPEC-wide policy consensus. In later years this was bolstered by specific resolutions of support for special cases.[36] Not until the 1967 war was the absence of a policy consensus outside the limited sphere of oil pricing and member sovereignty over natural resources perceived as a problem.

The Crisis of 1967

The June 1967 war marked a crisis point in Arab politics. The escalation of conflict to the violence of combat between Arabs and Israelis was only one aspect of it. Another, critical to Arab self-esteem and to intra-Arab relations, was the decision to use the "oil weapon" against those nations thought to be aiding Israel.[37] In response to Egyptian charges accusing the United States and Great Britain of aiding Israel militarily, Arab oil-exporting nations began an embargo of oil supplies to the United States. Some of these nations included Great Britain in their embargoes and a few denied oil to West Germany as well. Imperfections in the implementation of the embargo policy appeared to be due to actions of individual governments or to labor strikes, such as the one in Libya.[38] Only Saudi Arabia and Kuwait had sufficient financial reserves to withstand shutdown for more than sixty days, but even in these countries complete shutdown of installations was quickly reversed.[39] By 14 June the embargo was in place for all the Arab oil exporters except Libya and Iraq, although variations in the target nations still remained.[40]

In spite of increases in oil prices, due mainly to a fourfold increase in transportation costs for moving oil the longer distance around Africa rather than through the Suez Canal, neither the United States nor nations in Europe experienced an actual shortage of oil. There were several reasons for this. First, after the 1956 Suez crisis, oil-importing nations had laid in sizable reserve stocks of oil. It was estimated that Europe could have held out for six months on a combination of its reserves plus increased production from existing capacity in nonbelligerent nations.[41] Second, nonbelligerent nations such as Iran and Venezuela were able and willing to increase their production. In August 1967, these countries produced an additional 850,000 barrels of petroleum per day.[42] Third, the United States, by using reserve stocks and increasing domestic production, not only satisfied its domestic needs but was able to ship to Europe an additional 238,000 barrels of oil per day.[43] By 24 July an industry publication was able to report that non-Communist oil production had returned to its prewar level, owing primarily to production increases in the United States and Venezuela.[44]

Meanwhile, the production cutbacks and embargo were cutting deeply into the financial positions of Arab countries. Saudi Arabia's oil minister, Sheikh Ahmad Zaki Yamani, estimated that his country had lost $21 million from 5 June to 29 June alone, and he said he believed the producer governments were hurting themselves more than the targets of the embargo.[45] The Arab governments were torn by the need to finance the repair of damages left by the war and the simultaneous desire to withhold oil from unfriendly states, a policy that reduced their financial capabilities. By August 1967, it was clear that the partial embargo "had failed to produce the desired political or economic results."[46]

From the many proposals discussed among Arab officials at a series of meetings in the summer of 1967 emerged two distinct approaches to the problem of the failed partial embargo. Iraq offered the more radical plan. It proposed a total embargo on all Arab oil exports for a period of three months, to be followed by a partial embargo involving those states that had participated in or expressed sympathy with the aims of "Zionist and imperialist aggression." The total embargo was to involve the shutdown of pipelines running through Arab territory and continued closure of the Suez Canal. A number of financial and commercial moves directed against Western oil-consuming nations were to be made simultaneously. These would have reduced Arab-Western commericial, financial, and cultural ties by reducing Arab dependence on the West.[47] The Iraqi plan included proposals directed at dealing with the anticipated shortfall in foreign exchange resulting from implementing the embargo, and there was an additional proposal to establish a joint Arab fund for financing economic development projects in which states would participate according to their respective financial resources.[48]

The portion of the plan calling for the establishment of a development fund and the spirit of that portion concerned with lessening Arab dependency on the West met with general agreement. However, the embargo proposals were "reported to have been favored by the UAR, Syria and Algeria (with reservations, particularly concerning oil supplies to France) but to have met with general opposition elsewhere, *particularly from the other oil producing countries.*"[49]

These countries, principally Kuwait, Saudi Arabia, and Libya, favored a more moderate approach. Oil shipments should be resumed. Substantial funds accruing from oil profits and increased taxes in the more prosperous Arab countries should be directed to rebuilding the war-ravaged Arab states and increasing Arab military capability. They too wanted a stronger Arab policy with regard to obtaining higher oil prices and more Arab participation in the oil industry, but their focus was on negotiation rather than coercion as the preferred method by which to achieve these goals. The moderates also wanted to establish an Arab petroleum organization whose membership would be restricted to those Arab states "in which petroleum exports constitute a major source of revenue."[50] This differed significantly from previous proposals advocating such an organization that would have permitted non-oil-producing Arab states to join as well. In the past and during the 1967 crisis the non-oil-producers were usually the most vociferous in regard to calling for the use of the oil weapon. The 1967 embargo had seriously crippled the economies of the large producers. An Arab petroleum organization restricted to those countries whose economies depend most heavily on oil revenues would place the decision on whether or not to use the oil weapon in the control of those upon whom the most severe consequences would fall.

The key to the moderate "victory" at the Khartoum Conference of the Heads of Arab States (30 August to 1 September 1967) was the switch by Nasser from the radical to the moderate side. His move was very likely prompted by the military and financial exhaustion of the United Arab Republic and its need for financial aid from moderate oil-exporting countries, aid that would be forthcoming only if the embargo were lifted. The Nasser-moderate alliance was signaled by a concurrent bilateral pact between the UAR and Saudi Arabia, in which Nasser agreed to withdraw Egyptian troops from Yemen. In return for support in ending the selective embargo, Kuwait, Libya, and Saudi Arabia agreed to fund postwar reconstruction in Egypt and Jordan.[51] The partial embargo was lifted in September and Middle Eastern oil patterns quickly returned to normal except for altered transport arrangements caused by the continued shutdown of the Suez Canal.[52]

Immediately prior to the conference, on 14 September, Saudi Arabia had accepted an Aramco offer to eliminate completely the allowance granted the companies under the royalty expensing agreements as a precondition of reopening the Tapline (Trans Arabian Pipeline). At a consultative meeting in Taif following the Rome OPEC Conference meeting, Middle Eastern members of OPEC nominated Saudi Arabia as their representative in negotiations with the oil companies to eliminate a 6.5 percent discount the companies then received on their taxes. The companies owning Aramco (Esso, Texaco, Mobil, and Chevron) made an offer that would have eliminated the discount over a period of years, but no agreement was reached before the OPEC Conference met again in November. Full-scale country-company negotiations were opened in Teheran on 18 December, negotiations that OPEC members envisaged as "final" with regard to the allowance issue. But the negotiations failed to result in a formal agreement on any of the issues discussed. Even the countries that considered themselves moderate in their approach to country-company relations were angry at what they viewed as company obduracy on relatively trivial financial matters.[53]

By the end of 1967 all of the OPEC countries shared, to some degree, Arab anxiety concerning national autonomy in oil matters. Iran, Indonesia, and Venezuela feared that the oil price increases they had enjoyed during the supply disruption would melt away as Arab production resumed. The Arab nations, especially those committed to extensive aid programs for postwar reconstruction, were also committed to increasing their price per barrel of exported oil in order to be able to finance their aid programs and domestic economic development projects. Iran and the other non-Arab members of OPEC were solidly united in favor of moves to increase oil prices.

Resolutions passed at various Arab conferences in the summer of 1967 showed broad agreement within the Arab community favoring policies designed to increase host country influence on oil company price decisions. In spite of the agreement, the major Arab oil-exporting nations that favored moderate or long-term methods to achieve this influence felt that they had to resist encroachments on their autonomy not only by the companies but by the non-oil-exporting Arab states as well. The year's events had demonstrated a lack of appreciation on the part of non-oil states for the internal consequences suffered by the major exporters as a result of production cutbacks. Had the Iraqi proposal of a complete shutdown of oil production for three months been adopted, in addition to suffering financial sacrifices the entire country of Kuwait would have been virtually shut down as well, owing to a lack of natural gas (produced in conjunction with Kuwait's oil).[54] The major Arab oil-exporting countries were anxious to have a vehicle that would allow them to formulate collective Arab petroleum policy while affording them institutional protection against other Arab states whose dependence on oil was insignificant or nonexistent. Because both Arab and non-Arab members of OPEC were solidly behind efforts to increase oil prices, it was not intended that the new organization impair the effectiveness of OPEC nor change in any way Arab participation in OPEC. With these aims in mind, its founders established the Organization of Arab Petroleum Exporting Countries—OAPEC.

OAPEC and Its Critics

Libya, Kuwait, and Saudi Arabia participated in the signing of the OAPEC Agreement on 9 January 1968. No representative of Iraq attended the ceremony because Iraq had chosen not to affiliate with the new organization. There were three main reasons for the split between Iraq and the others. First, Iraq was displeased by the exclusiveness of OAPEC, believing that it simply created another axis along which intra-Arab division might occur. Iraq preferred that membership in OAPEC be open to any Arab oil-producing state, but the others had limited OAPEC membership to those for whom oil was the principle source of national income. Had the Iraqi point of view prevailed, OAPEC would have had a potential membership closer to that of the Arab League and the major exporters would have found themselves in another organization where their economic preferences would have been in the minority.

Second, Iraq was reportedly at odds with OAPEC because of the degree of independence OAPEC enjoyed from its member states, particularly as reflected in the provision for an OAPEC court to arbitrate disputes among the members.[55]

A third reason for Iraqi opposition to OAPEC was injured pride. Independent sources charged that the decision to locate the OAPEC headquarters in Kuwait rather than in Baghdad was a significant factor in Iraq's refusal to join.[56]

The defection of Iraq was a matter of some concern to the other members. As the only major Arab oil exporter not within the OAPEC fold, socialist Iraq's nonparticipation in an oil producers' organization in which all the other member governments were monarchies (or for that matter, the nonparticipation of radical Iraq in an oil producers' organization in which all the other members were moderate in their positions toward the oil companies) was a reminder of ideological splits in the Arab world. Although the other members of OAPEC refused to regard Iraq's position as either ideological or permanent,[57] OAPEC without Iraq made the new organization seem more sinister to those Arabs whose primary policy concern was Arab unity rather than petroleum. Objections were raised in the Kuwait Assembly during the debate on ratification of the OAPEC Agreement because it excluded some oil-producing Arab countries from membership.[58] Similar criticism came from outside OAPEC, in one instance from Nicola Sarkis, the Lebanese economist and petroleum consultant. He felt that the transit countries, those nations such as Syria through which oil pipelines passed, were threatened by oil company policy as much as the major Arab oil exporters. Even before the 1967 closure of the Suez Canal, oil companies as well as private entrepreneurs had begun investing in supertankers to transport oil. According to Sarkis, this happened because the companies were becoming less and less able to "appoint and dismiss governments, organize military coups..." and otherwise manipulate transit country governments.[59] OAPEC's exclusionary policy thus denied the transit countries a lever in their efforts to command a share in the economic benefits of Arab oil. Sarkis also felt that the nature of the cooperation among OAPEC members committed them to a policy of negotiation with the oil companies to improve concession terms, an approach he regarded as futile and an attempt to delude the Arab public. In addition, Sarkis predicted that OAPEC would be redundant because OPEC was dealing adequately with problems OAPEC might want to address.[60]

In contrast to Sarkis's view that OAPEC would ultimately serve the interests of oil companies rather than those of oil exporters were reports in the Kuwait press that the oil companies themselves were strong opponents of OAPEC. When OAPEC Acting Secretary General Sheikh Ahmad Zaki Yamani was asked by a Kuwaiti correspondent for comments about this source of opposition to OAPEC, he chose to interpret it in the context of OAPEC's commitment to participation.[61] Yamani asserted that participation was "an established reality" unlike nationalization, which he called "an empty slogan." He used the occasion to answer critics like Sarkis, who believed that OAPEC was too soft to deal with the oil companies, by construing company opposition as a testimonial to OAPEC's efficacy in challenging the companies' control of the oil industry in the Middle East.[62]

During its first year, OAPEC officials made frequent public statements designed to answer critics and to allay general fears that OAPEC had been intended to benefit the moderate major Arab oil exporters exclusively. For example, OAPEC leaders proposed to cooperate with other formally organized Arab Petroleum groups, refusing to put themselves in the position of rival to other efforts to bring about intra-Arab cooperation.[63] OAPEC offered formal support to OPEC-sponsored subgroups such as a proposed "petrochemical OPEC" to include Iran, Kuwait, and Saudi Arabia[64] and invited non-OAPEC participation in OAPEC-sponsored projects.[65] Significantly, this welcome was extended not just to other Arab nations or to other petroleum-exporting nations but to the major oil companies and to the "independents" (oil companies that are not fully integrated vertically from production to marketing) as well. In addition, the OAPEC Agreement provides for non-members to attend OAPEC meetings by invitation.[66] Indeed, Iran was represented during OAPEC deliberations that led to the decision to institute a fourth Arab oil embargo in October 1973.[67]

The criticism directed at OAPEC's membership policy and to its unusual degree of institutional autonomy was fundamental. OAPEC was a major departure from the tradition of consensus politics in the international organizations in which Arab states were members. It was probably only a matter of time before the issue of the nature of OAPEC had to be faced by all Arab nations, members and nonmembers, and resolved. This did not happen until Iraq applied to become a member of OAPEC in 1970. Iraq then became the focal point of a struggle between moderate and radical members of OAPEC over the goals of the organization and the distribution of power within it. In effect, this struggle was an identity crisis for OAPEC as an organization. It began in the fall of 1969 when radical elements in Libya overthrew the Idris monarchy and set up a revolutionary regime. The new regime was much more vigorous than the old regarding oil policy. Although Libyan Oil Minister Izz al-Din al-Mabruk ruled out nationalization of the country's oil industry as an official policy of the new regime, Libya soon heated up ongoing efforts to increase posted prices for its oil. It also launched a campaign to increase Arab participation in OAPEC. Libya implemented its new policy toward OAPEC by broadening an existing petroleum cooperation agreement it had with Algeria and by extending the agreement to Iraq and the United Arab Republic as well in January 1970.

When the OAPEC Council met in January 1970, it responded to the Libyan initiatives by appointing Libyan Suhail Sa'dawi secretary general of OAPEC and by voting to support Libya in its scheduled negotiations with the oil companies on prices.[68] OAPEC also retreated from a previous policy whereby it participated as a single entity within Arab League petroleum councils. Resolution IV.10[69] of the OAPEC Council provided that each member state would participate individually at oil meetings of the Arab League, although after the members had met beforehand to coordinate their national positions. Prior to the passage of Resolution IV.10, OAPEC members had sent a single delegation to these meetings. In this way Libyan influence began to erode

OAPEC autonomy and to increase member autonomy. The weight of OAPEC within Arab League oil policymaking bodies was decreased.

Shortly after the January meeting, Libya sponsored radical Algeria for membership in OAPEC. Under a liberal interpretation of Article 7 of the OAPEC Agreement, Algeria was accepted as a member of OAPEC in May 1970 along with four oil-exporting gulf emirates: Qatar, Abu Dhabi, Bahrain, and Dubai. Concurrently Libya and Algeria strengthened petroleum ties with Iraq via a formal agreement among the oil ministers of the three countries. This relationship stressed the role of revolutionary governments in "recovering the sources of national wealth and placing them in the service of the people's interests."[70]

All this served to increase radical state influence within OAPEC, directly by Libya and Algeria, and indirectly through these nations' petroleum alliances with nonmember radical regimes. In addition, the new energy displayed by the radical states in pressing for rapid changes in the oil industry in the Middle East represented a threat to the moderates' commitment to gradual change. When Iraq applied for membership in OAPEC shortly before the June 1970 meeting of the OAPEC Council, the council responded by adjourning without acting on the application. Clearly a threshold in moderate tolerance had been reached.

That adjournment marked the beginning of a year of delay and postponement on Iraq's application. The June 1971 council meeting broke up after only two days, with OAPEC on the verge of schism over the issue of Iraq. By then OAPEC was split between Algeria, Libya, and Kuwait, which favored the admission of Iraq, and Saudi Arabia and the gulf emirates, which were opposed. As the delays continued into fall 1971, pressure on those opposed to the admission of Iraq came from within the moderate coalition as Abu Dhabi defected to support Iraq. Shortly after this change, Libyan Oil Minister Mabruk warned that OAPEC would have to be dissolved if the council could not resolve the issue before much longer. Mabruk extended his criticism of OAPEC to what he regarded as an almost dilettantist approach to its economic projects as well. "Had the Organization, which is four years old, taken its work seriously, it would have been able to realize many important projects which are lacking in the Arab World today."[71] Mabruk thought that the organizational paralysis over the issue of Iraq's membership was symptomatic of OAPEC's overinvolvement with political issues at the expense of attention to the commercial goals responsible, in his view, for its establishment.[72]

Mabruk's idea of OAPEC as a strictly commercial venture reflected the results of all the changes in the organization since the Libyan coup as much as it represented politicking on his part to legitimize the changes by inplying they had not occurred. But although the founding members had professed a basically economic orientation for OAPEC, early policies, such as unified

OAPEC representation in the Arab League, were indicative of OAPEC's reach for a wider political role. It would have been naive for either moderates or radicals actually to believe that an organization like OAPEC could ever be apolitical. The rash of cooperative petroleum agreements among radical states following OAPEC's founding[73] mark recognition of the potential political threat OAPEC posed to other Arab states. But when Libya's regime changed, it forced a reduction in OAPEC's political role within the Arab League because the Libyan revolution created disunity within OAPEC. This dampened whatever effect an "oil bloc" might have had on Arab League politics. It also pushed OAPEC to support more rapid devolution of control of oil to the producing states than the moderate nations were ready for. Both of these are essentially political positions, but the former involves a conflict over centralized versus dispersed power, whereas the latter assumes dispersed power and recalls OPEC's device of limited consensus and permissive implementation by concentrating on substance rather than on process. The Libyan coup had occurred too early in OAPEC's history for the organization to have developed a tradition of independence to match the potential written into its charter. Survival in the face of drastically opposed regimes and their concomitant philosophies regarding centralized and decentralized power required the suppression of political activity on issues of process. The nearly schismatic response to the issue of admitting Iraq was less concerned with Iraq per se than with whether OAPEC would survive as an organization producing substantive economic benefits or die as a process mechanism for integrating the petroleum policymaking organs of its member governments.

As it happened this decision belonged to Saudi Arabia, which chose to let OAPEC live in anticipation of its potential to rationalize important facets of the economic development of the region.[74] Symbolically, Iraq was admitted, along with Egypt and Syria, under a Saudi-sponsored amendment to Article 7, the membership article of the OAPEC Agreement.

The Triumph of Substance

Proponents of community in the Middle East can still applaud the Saudi Arabian decision to compromise on the nature of OAPEC because a live organization is a better vehicle for cooperation than either a dead one or a set of surviving antagonistic fragments. Centralizing political power in a strong international organization is one avenue toward community but perhaps not the best one when basic agreement on the proper dispersion of power between the members and the community is impossible. Structural community among nations can also arise from an economic joint venture approach that results in functional interdependence among participants.[75] Shared facilities, running the gamut from an oil refinery constructed to run on a mixture of crudes of international origin to rules permitting the free movement of workers across international borders, creates connections among nations whose rupture would involve high costs to the individual countries. These interdependencies or transnational connections force nations to moderate conflict among themselves in order that the functional linkages not be broken and the high costs incurred. Positively, the linkages contribute to a higher total level of welfare for the linked nations. This comes from actual cost savings on shared facilities and also from increased opportunities for a more productive use of national resources such as labor, capital, or petroleum within the security of community arrangements.

Summary and Conclusions

All complex organizations must face the problem of maintaining a consistent policy stance while preserving some flexibility in the way they react to specific situations. An international organization has an even crueler problem in that its members are autonomous. Their participation in the international organization is only a minor and often an unimportant aspect of their identities. An international organization, unlike a corporate conglomerate, is always faced with the likelihood that a member will refuse to implement a collective policy or even leave the organization if it disagrees violently enough. Specific policy alternatives or entire issue areas may be foreclosed to the international organization regardless of its statutory mandate because the probability of conflict greatly outweighs the likelihood of cohesion.

The tension between conflict and cohesion becomes greater as the international organization increases in size. New members introduce additional diversity that has to be accommodated in subsequent policy decisions.[76] Policy choices are increasingly constricted and the variety of implementation strategies potentially increases. Situations demanding a clear policy and uniform implementation in order to be effective become explosive when the interests of the members of an international organization diverge.

Neither the Arab League nor OPEC could deal effectively with the simultaneous political and economic dimensions of oil policy for their Arab oil-exporting members. The Arab League was established to coordinate Arab foreign policy, to foster Arab cooperation, and to protect Arab sovereignty from outside encroachment. Politically, the Arab League examined ways to use petroleum as a weapon in the conflict with Israel, while technically it provided for the exchange of information through its Committee of Oil Experts and Arab Petroleum Congresses. In choosing to include non-Arab oil-exporting countries in the first Arab Petroleum Congress in 1959, the Arab League provided the seedbed that produced OPEC. But it nearly accomplished OPEC's destruction as well when its 1967 oil embargo was met by increasing oil production in non-Arab OPEC members. This economic opportunism could have split OPEC had its members not been able to agree on a militant pricing policy stance against the oil companies once the embargo had ended.

OPEC remedied a serious deficiency in any exclusively Arab petroleum organization with economic goals: it included all significant Third World exporters of oil. Over time OPEC rather than the Arab League became the main vehicle for the coordination of Arab oil policy. First the United Arab Republic and subsequently other Arab states proposed the creation of an Arab petroleum organization to act as a counterweight to OPEC. As originally proposed, any Arab nation would have been allowed to belong to the Arab Petroleum Organization, in effect re-creating a small-scale Arab League that would oversee only oil policy. The major producers opposed this strenuously, some because they disliked the idea of leaving their biggest industry to the mercy of the political whims of non-oil-exporting nations. Arab oil consultants Abdullah Tariki and Nicola Sarkis were also recorded as opposing an Arab petroleum organization on the grounds that, as a quasi-Arab League, it would share the other league's inability to enforce any of its decisions. These critics thought that an Arab petroleum organization would give the oil companies an opportunity to divide OPEC by forcing it to cope with irrelevant political issues. Their reply to the resolution of the Baghdad Conference calling for the creation of an Arab petroleum organization was a suggestion to strengthen OPEC.

OPEC, however, was not able to cope with the peculiar problems of its Arab members. However successful it might be at preserving and increasing host country autonomy in relationships with the oil companies, OPEC could neither shield the major Arab oil-exporting countries against the policy demands of non-oil Arab states nor prevent its non-Arab members from taking economic advantage of any future Arab oil embargo. The Arab oil exporters remained uniquely vulnerable as a group in both organizations. They responded by proposing the creation of an Arab petroleum organization from which non-oil states would be excluded. Their new organization was to be able to represent collectively the interests of Arab oil-exporting countries as neither other organization had been able to do.

OAPEC was founded in January 1968 as a reflection of the ideology and goals of the moderate Arab oil-exporting nations. Its charter provided for a strong central authority, exemplified most clearly in its exclusive membership policy and in the provision for a judicial board with authority to decide intra-OAPEC disputes. Demonstrating member support of OAPEC's centralized character, during its first two years OAPEC attended oil meetings of the Arab League as a single entity representing its members. But when the coup in Libya resulted in the introduction of a radical republican regime into

OAPEC policy meetings, internal consensus on the identity and purposes of OAPEC was destroyed. An institutional identity crisis resulted when Iraq applied for membership in OAPEC. The charter requires that all three founding members approve each applicant, but Saudi Arabia was displeased with the changes in OAPEC already accomplished by radical members and refused to accept Iraq. These changes involved reducing the OAPEC representative at Arab League deliberations to observer status, while individual government representatives resumed their pre-OAPEC roles. Substantively, Libya demanded OAPEC support for its pricing and production limiting goals and campaigned to open OAPEC membership to all Arab oil producers. Algeria was admitted to OAPEC under Libya's sponsorship despite its failure to conform to the letter of the charter-mandated membership requirements. Algeria then demanded that OAPEC support its efforts to gain control over its petroleum, despite the fact that it acted at least once without even the pretense of prior consultation with the other members.[77]

In addition, both Libya and Algeria sponsored and participated in rival petroleum organizations composed of revolutionary-regime nations. Continued delay over the Iraqi application for membership increased the likelihood that OAPEC would disintegrate and one of these radical unions would become a major Arab petroleum organization whether or not Saudi Arabia and its few supporters could maintain a rump organization of moderates. In the face of actual schism, Saudi Arabia relented and sponsored an amendment to Article 7, the membership article, of the OAPEC Agreement. The amendment permitted membership in OAPEC to virtually all Arab oil-exporting countries. This marked the end of the OAPEC attempt to integrate the major Arab oil exporters under a strong central policymaking authority and a tacit return to the principal of Arab unity in petroleum policymaking.

The changes begun in 1969 gradually pushed OAPEC into becoming more and more like the Arab Petroleum Organization that had been proposed originally. That organization was to have permitted membership to any Arab state. Internally it would mirror the conflicts of the Arab League and be a political threat to no other Arab organization through any potential to divide the Arab countries over oil policy. By 1973 the transition was almost complete. At the request of the Arab League OAPEC itself imposed an oil embargo against the friends of Israel during yet another Arab-Israeli war.[78] In effect OAPEC answered its Arab critics by becoming the sort of organization they had wanted all along.

Notes

1. George Lenczowski, "Arab Radicalism: Problems and Prospects," *Current History* 60 (January 1971); Malcolm Kerr, "Regional Arab Politics and the Conflict

with Israel," in Paul Hammond and Sidney Alexander, eds., *Political Dynamics in the Middle East* (New York, 1972).

2. Kerr, "Regional Arab Politics"; M. F. Anabtawi, *Arab Unity in Terms of Law* (The Hague, 1963); A. S. Becker, B. Hansen, and M. H. Kerr, *The Economics and Politics of the Middle East* (New York, 1975), Chap. 2.

3. Robert W. MacDonald, *The League of Arab States* (Princeton, 1965), pp. 37-38.

4. Anabtawi, *Arab Unity in Terms of Law,* Chap. 1.

5. MacDonald, *League of Arab States,* p. 40.

6. Ibid., p. 41.

7. Anabtawi, *Arab Unity in Terms of Law,* p. 70.

8. MacDonald, *League of Arab States,* pp. 37-38.

9. Ibid., p. 292.

10. Ibid., pp. 249-63; Phillippe Schmitter, *Autonomy or Dependence as Regional Integration Outcome: Central America,* Institute of International Studies Research Series, No. 17 (Berkeley, 1972), p. 61. Schmitter uses Arab cohesion as a standard and compares cohesion among other groups to it.

11. Donald L. Losman, "The Arab Boycott of Israel," *International Journal of Middle East Studies* 3 (April 1972).

12. MacDonald, *The League of Arab States,* p. 284.

13. On 31 March 1979, nineteen members of the Arab League (including Palestine) passed a series of resolutions condemning Egypt for signing a separate peace treaty with Israel. These resolutions involved both political and economic sanctions, including the suspension of Egypt's membership in the Arab League and the planned severance of diplomatic ties between Egypt and the eighteen nation-state members adopting the resolutions (see *The New York Times,* 1 April 1979, pp. 1, 4). Egypt suspended its own membership in the Arab League at one point under Nasser. Suspension by a country itself or by its co-members is serious, but it is less so than an absolute severance of the state from the organization because suspension implies eventual reinstatement. Egypt is also currently suspended from OAPEC and scores of other Arab groups for the same reason. See below.

14. R. E. Thoman, "Iraq and the Persian Gulf Region," *Current History* 60 (January 1973), discusses the role of the Arab League in persuading Iraq to relinquish its claim to Kuwait after Kuwait became independent of Britain. For a discussion of the issues of Iraq's claim, see Husan M. Albaharna, *The Arabian Gulf States,* 2d rev. ed. (Beirut, 1975), Chap. 15.

15. Losman, "The Arab Boycott of Israel"; also Zuhayr Mikdashi, *The Community of Oil Exporting Countries* (Ithaca, 1972), Chap. 1.

16. Mikdashi, *Community of Oil Exporting Countries,* p. 28.

17. Fuad Rouhani, *A History of O.P.E.C.* (New York, 1971), p. 76.

18. G. and H. Stevens, "The First Arab Petroleum Congress," *The World Today* 15 (June 1959).

19. One scholar has reported that there was a strong feeling that the companies had lowered posted prices to retaliate against the congress and its implication of host country collusion. See David Hirst, *Oil and Public Opinion in the Middle East* (New York, 1968), pp. 43-44.

20. Zuhayr Mikdashi, *A Financial Analysis of Middle Eastern Oil Concessions, 1901-1965* (New York, 1966), p. 172, Table 19.

21. Mikdashi, *Community of Oil Exporting Countries,* pp. 31-32.

22. Ibid.

23. *Oil and Gas Journal (OGJ),* 15 August 1960, pp. 83-85; Mikdashi, *Community of Oil Exporting Countries,* p. 33.

24. Organization of Petroleum Exporting Countries (OPEC), OPEC Statute (Vienna, 1965); and Resolution I.1 of the OPEC Conference. (OPEC Resolutions are numbered serially in Arabic numerals, and the OPEC Conferences are numbered serially in Roman numeral prefixes. Thus, Resolution I.1 was the first OPEC Resolution [1] and was passed at the first OPEC Conference [I].)

25. Mikdashi and Rouhani share this perspective as does Mana Saeed al-Otaiba, *OPEC and the Petroleum Industry* (New York, 1975); Fadhil Al-Chalabi, "Pricing of OPEC Crude Oil: A Case for the Valuation of Depletable Resources in Relation to Economic Development," mimeographed, (n. p., n. d.), obtained from OAPEC courtesy of Dr. George Tomeh; and many others.

26. Initially, OPEC members hoped to increase their incomes through increased production. See George Lenczowski, "The Oil Producing Countries," in Raymond Vernon, ed., *The Oil Crisis* (New York, 1976), p. 60. They grew more interested in encroaching on the rents of the companies as it became apparent that the elasticity of demand for oil was very low. See Richard C. Weisberg, *The Politics of Crude Oil Pricing in the Middle East, 1970-1975,* Research Series No.31 (Berkeley, 1977), p. 74. This change was reflected in the passage of OPEC Conference Resolution XVI.90 stating that excessively high net earnings after taxes by the oil companies would constitute grounds for renegotiating contracts.

27. Rouhani, *A History of O.P.E.C.,* p. 158; George W. Stocking, *Middle East Oil: A Study in Political and Economic Controversy* (London, 1970), pp. 350-60; Naiem Sherbiny, "Arab Oil Production Policies in the Context of International Conflicts," in Naiem A. Sherbiny and Mark A. Tessler, eds., *Arab Oil: Impact on the Arab Countries and Global Implications* (New York, 1976), p. 42.

28. Expensing rather than deducting royalties increases government revenue per barrel of oil. For example, at a posted price of 100 units, costs of 10 units, royalties of 15 units, and a tax rate of 50 percent, royalty deduction results in government income of 45 units per barrel, while royalty expensing gives the government 52.5 units of income per barrel.

29. Rouhani, *A History of O.P.E.C.,* p. 203.

30. Edith Penrose, "OPEC and the Changing Structure of the International Petroleum Industry," in Zuhayr Mikdashi, S. Cleland and I. Seymour, eds., *Continuity and Change in the World Oil Industry* (Beirut, 1970), Mikdashi regards OPEC's ability to increase dollar prices for oil as less significant because the increases "were largely offset by increases in the price levels" of OPEC member imports (Mikdashi, "The OPEC Process," p. 203). But, as Penrose points out, OPEC managed to increase oil prices in the face of an oversupply of oil in the world market.

31. Rouhani, *A History of O.P.E.C.,* pp. 217-33; Weisberg, *The Politics of Crude Oil Pricing in the Middle East,* pp. 23-31.

32. Rouhani, *A History of O.P.E.C.,* pp. 233-35.

33. Ibid., pp. 245-50.

34. Ibid., pp. 229-33, quote on p. 233; Weisberg, *The Politics of Crude Oil Pricing in the Middle East*, pp. 27-31.

35. U.S. Congress, Senate, Committee on Foreign Relations, Subcommittee on Multinational Corporations, *Multinational Corporations and United States Foreign Policy*, Pt.5, 93rd Cong., 1st sess., 1975, pp. 104-14, 196-97, 213. Weisberg believes that the shift in power from the oil companies to the producing governments began with the success of the royalty-expensing negotiations. See Weisberg, *Politics of Crude Oil Pricing*, p. 33.

36. Rouhani, *A History of O.P.E.C.*, pp. 17, 30.

37. The "oil weapon" is the use of oil policy to achieve foreign policy goals. The term is most frequently associated with anti-Israel policy and has usually meant an oil embargo. Farouk Sankari, "The Character and Impact of Arab Oil Embargoes," in N. Sherbiny and M. Tessler, eds., *Arab Oil*.

38. *Petroleum Press Service (PPS)*, July 1967, p. 247.

39. Ibid.; *OGJ*, 19 June 1967, pp. 76-77.

40. Onnie Lattu, "Remarks," delivered before the American Petroleum Institute Southern District Division of Production in San Antonio, Texas, 7 March 1968, p. 4. The author was the director of the Oil and Gas Office of the Department of the Interior at the time.

41. Ibid., p. 3; *PPS*, July 1967, pp. 244-45.

42. Lattu, "Remarks," p. 4.

43. Ibid.

44. *OGJ*, 24 July 1967, p. 30.

45. *OGJ*, 10 July 1967, p. 96; *The Middle East Economic Survey (MEES)*, 11 August 1967, pp. 5-8.

46. *MEES*, 12 January 1968, p. 3.

47. *MEES*, 11 August 1967, pp. 5-8. This plan involved the nationalization and "Arabization" of "monopolistic foreign economic interests plus all banks, insurance companies and commercial enterprises which try to thwart the Arab boycott." One aim of this Draconian proposal was to force the Arab countries to become more self-sufficient.

48. *MEES*, 11 August 1967, pp. 5-8; *OGJ*, 28 August 1967, p. 96.

49. *MEES*, 25 August 1967, p. 2, emphasis added.

50. *MEES*, 25 August 1967, p. 2.

51. *MEES*, 1 September 1967, pp. 1-2; *MEES*, 8 September 1967, p. 4; *OGJ*, 4 September 1967, pp. 1-2; George Lenczowski, "Arab Bloc Realignments," *Current History* 52 (December 1967).

52. *OGJ*, 25 September 1967, p. 90; *PPS*, September 1967, p. 343.

53. *MEES*, (22 December 1967,) p. 14.

54. *MEES*, "Supplement," 7 June 1968, p. 2.

55. *MEES*, 13 September 1968, p. 7.

56. *MEES*, 13 September 1968, p. 7; *MEES* 14 June 1968, p. 8.

57. *MEES*, 7 June 1968, interview with Sheikh Yamani. The interview is unpaged.

58. *MEES*, 14 June 1968, p. 7. Egypt and Syria were not admitted to OAPEC until Article 7 of the OAPEC Agreement was amended in 1972. The agreement will be discussed in detail in the next chapter.

59. *MEES*, 9 February 1968, p. 8.

60. *MEES*, 9 February 1968, p. 8.

61. The concept of participation was invented by Yamani. It means that oil-producing countries are given shares of concession companies, thus enabling them to share profits while leaving undisturbed downstream outlets for their oil. Participation was proposed as an alternative to nationalization which would preserve producing country oil markets. When Iran nationalized its oil companies in 1951, buyers of Iranian crude refused to purchase oil, cutting off Iran's chief source of income. Participation was designed to allow oil-producing countries to share profits without endangering income.

62. *MEES*, 15 November 1968, p. 12.

63. *MEES*, 15 November 1968, p. 123, *MEES*, 24 January 1969, p. 9.

64. *MEES*, 13 September 1968, p. 7.

65. *MEES*, 27 December 1968, p. 9.

66. OAPEC Agreement: Article 10. See Appendix 1 for the full text of the agreement.

67. Marwan Iskander, *The Arab Oil Question* (Beirut, 1974), Chap. 1.

68. *MEES*, 30 January 1970, pp. 7-8.

69. The system for numbering OAPEC Resolutions is like the OPEC system. The Roman numeral prefix denotes the number of the council that passed the resolution, and the Arabic numeral represents the order of specific resolutions, which are serially numbered.

70. *MEES*, 3 July 1970, p. 4.

71. *MEES*, 5 November 1971, p. 9.

72. Ibid.

73. *MEES*, 3 July 1970, p. 3. This pattern is similar to the rash of partial unification efforts that followed the establishment of the United Arab Republic in 1958 (see Chapter 1).

74. For example, by providing a vehicle for joint ventures, and a formal, institutional setting for coordinators of economic development projects in petroleum sponsored by the governments individually.

75. Alex Inkeles, "The Emerging Social Structure of the World," *World Politics* 27 (July 1975); Richard N. Cooper, *The Economics of Interdependence: Economic Policy in the Atlantic Community* (New York, 1968), Chap. 1.

76. Philip Selznick, "Foundations of the Theory of Organization," in Amitai Etzioni, ed., *Complex Organizations* (New York, 1965), p. 27.

77. *MEES*, 21 August 1971, p. 7.

78. The only published work I have seen that alludes to the role of the Arab League in the 1973 embargo decision is Ibrahim Shihata, *The Case for the Arab Oil Embargo* (Beirut, 1975). This was confirmed to me in a letter from George Tomeh (at that time an OAPEC consultant and former Syrian ambassador to the United Nations) dated April 3, 1978: "to attribute the rises in the price of oil in October, 1973 to OAPEC as such is wrong.... The original invitation for them [the Arab oil states] to meet—not as OAPEC—but as Arab oil exporters, came from the Arab League and then was sponsored by the Government of Kuwait." The purpose of this meeting was to settle on the use of the oil weapon.

Chapter 3

OAPEC: How It Works and What It Does

The OAPEC Agreement

The OAPEC Agreement bears a family resemblance to the charters of the Arab League and OPEC. They are similar with regard to the order of provisions, most procedural details, and the specifications for each organization's internal structure.[1] Differences exist in qualifications for membership, the degree of agreement required for binding decisions and the locus of responsibility for enforcing them, organizational autonomy with respect to member nations, and the type of institutions each establishes. Differences of this type indicate how secure members feel with what each perceives as the national interests of other members included under the jurisdiction of the international organization.

The Arab League is the most permeable[2] of the three organizations measured by qualifications for membership. "Any independent Arab State has the right to become a member of the League."[3] Neither the OPEC Statute nor the OAPEC Agreement talks about *rights* to membership. Rather, each poses requirements for membership with regard to the petitioner's degree of involvement in the petroleum industry, OPEC members needing "a substantial net export of crude petroleum"[4] and OAPEC members having to show that oil "constitutes the principal and basic source of [their] national income."[5] Compared with the Arab League, OAPEC is more exclusive. OAPEC's exclusivity prompted criticism from insiders as well as from the excluded. The later amendment of the OAPEC membership article was more cosmetic than substantive; it now permits any Arab country for which petroleum constitutes "a significant source of its national income" to join OAPEC.

In addition to the required degree of involvement in the oil industry, petitioners for membership in OPEC and OAPEC face the prospect of blackball by current members who must approve each new member by a three-fourths majority vote (including the votes of all founding members). However, this provision is seldom effective in preventing the admission of a country able to command a substantial majority in its favor. A good example is the case of Iraq's petition to join OAPEC. OAPEC members were divided on the issue; a majority supported Iraq. Had Saudi Arabia's opposition continued it could have destroyed OAPEC, but a compromise was worked

**Figure 2: *The Relationship Among OAPEC, OPEC, and the Arab
League***

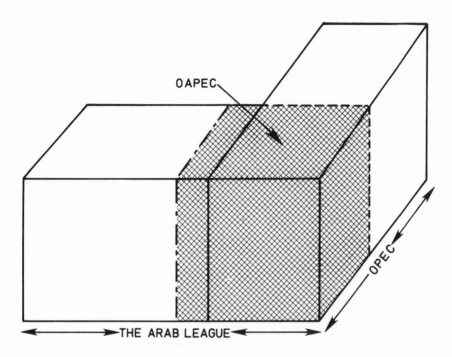

out that embraced more than this single issue. OAPEC survived, although
greatly changed. In effect, the three-fourths majority provision can moderate
but not prevent changes in membership that majorities in OPEC or OAPEC
favor.

Whether or not the requirements for membership in OAPEC are observed
to the letter, they have been able to limit membership to Arab countries
heavily dependent on petroleum revenues. Analytically, the universe of possible OAPEC members is limited to the Arab League, of which it is a subset.
Elements of the OAPEC subset are distinct from non-OAPEC elements in the
universe because of their dependence on petroleum revenues for a large part
of their national incomes. OAPEC members can be assumed to have interests in the economic health of the petroleum industry that are not shared by
members of the Arab League that do not export oil. In predicting the
likelihood of congruence in petroleum policy preferences among OAPEC
members or among Arab League members, one would expect those of the
former organization to coincide much more often.

OAPEC is the only one of the three organizations that has abandoned
consensus decisionmaking in favor of a modified form of majority rule. Both

the Arab League and OPEC require unanimous consent to make official policy. The Arab League allows two types of decisions: unanimous decisions, which are "binding upon all member states of the League," and majority decisions, which bind "only . . . those states which have accepted them."[6] This favors member state autonomy at the expense of the Arab League. Member state autonomy is further strengthened by the second clause of Article 7 of the Pact of the League of Arab States, which specifically omits a mechanism for the league to enforce its decisions. Enforcement is left "to each member state according to its respective laws" and dilutes the implementation of league policy, especially in member states whose support for specific policies might be less than enthusiastic.[7]

OPEC decisions are also taken unanimously[8] reflecting a similar preference for member autonomy over central authority. The reason for this can be traced back to the experiences of OPEC member countries in the 1950s and 1960s when one or another of them tried to increase its own share of the crude oil export market at the expense of the others. In practice, OPEC decisionmaking is more flexible than the statute would make it seem due to the organization's permissive attitude toward individual member behavior in some policy areas where only limited consensus can be achieved. But this strategy is ineffective in policy areas such as production prorationing.[9] A production prorationing policy may be impossible to achieve under a unanimous decision rule particularly in the absence of enforcement capability (see Chapter 5). OPEC does have within the secretariat an enforcement department that is charged with reporting on progress in member-country implementation of OPEC resolutions and recommendations.[10] But there is no way to force a recalcitrant member to comply when no progress takes place.

OAPEC stands in contrast to the Arab League and to OPEC both with regard to the degree of unanimity required to produce a binding decision and in its power to enforce decisions once they are made. Like the Arab League and OPEC, OAPEC allows each member nation a single vote, but the largest majority ever required for a binding decision is "three-fourths of the vote of the total membership, including those of at least two founding members."[11] A three-fourths majority is required for passage of "resolutions on substantive matters."[12] Procedural matters require only a simple majority vote[13] and the decision as to whether a matter before the council is substantive or procedural is also taken by a simple majority vote.[14] In effect, any matter a simple majority can couch in procedural terms can be passed by a simple majority vote.

Although Article 21 of the OAPEC Agreement provides that statutes and resolutions "shall be subject to ratification by the competent authorities in the member countries according to the legal rules in force," it also provides for the collective indemnification of a member injured by another's noncompliance with a binding resolution. This encourages ratification and similar levels of implementation of OAPEC resolutions on the part of all the members in order to avoid financial penalties. In addition, the OAPEC Agreement provides for a judicial body competent to initiate some types of proceedings and to hear other cases subject to the approval of one or more parties to a dispute. We will return to the matter of the OAPEC Judicial Board below.

OAPEC's relaxation of the unanimity requirement reflects the expectation of the founding members that a greater community of interests would exist among themselves than exists in either the Arab League or in OPEC. This expectation was reasonable given the limited scope of the organization and membership requirements designed to bar nations whose interest in petroleum policy was more political than economic. Limiting OAPEC to Arabs reflects the special responsibility Arab nations feel concerning the Palestine situation, which might obligate members in some instances to put politics above economics.[15] By limiting membership to major oil exporters and subsequently to nations receiving significant national income from petroleum, OAPEC sought to prevent a repetition of the events of 1967. Then the political decision to embargo had been taken by a body a majority of whose members were only minimally affected by the policy's economic consequences. The OAPEC process was designed to allow the major Arab oil exporters to make their own tradeoffs between politics and economics, rather than to be traded off by non-oil-dependent states whose ends they might agree with but whose choice of means might require sacrifices that would be unequally borne. OAPEC was expected by its founders to serve "to some extent as a substitute for the various oil activities of the Arab League,"[16] but with changes in the internal balance of interests such that Arab oil-dependent states rather than Arab states would be making oil policy.

OAPEC is potentially more independent of its member states than either OPEC or the Arab League. This is shown by provisions entitling OAPEC to enforce some of its decisions by applying coercion to nonconforming members. The legitimate exercise of coercive power is one of the attributes of sovereignty.[17] Of the three organizations, only OAPEC has been allowed by its members to exercise coercive power, even though the range within which it is permitted to act is very narrow.[18] Although it is limited, OAPEC autonomy still represents a surrender of member sovereignty in favor of collective decisionmaking.

As a mechanism for realizing its autonomy, OAPEC's charter provides for an organ not mentioned in the charters of the Arab League or of OPEC. This is the OAPEC Judicial Board (Court), the main provisions for which are outlined in Articles 21-25 of the agreement. These include the definition of the board's jurisdiction and a statement of the requirements for appointment to its bench. Judges are to be appointed by the council and must have international reputations as jurists, a necessary requirement for individuals serving on an international court. The board is authorized to decide disputes "relating to the interpretation and application" of the OAPEC Agreement and "the implementation of the obligations arising from it"; disputes "in the field of petroleum" which might arise among OAPEC members; and "disputes which the Council decides that the board is competent to consider."[19] This last provision is something of an "elastic clause" that will permit OAPEC to enlarge the jurisdiction of the board should time and circumstances change. The decision of only one party to any of the types of disputes already listed will be sufficient to give the board jurisdiction to hear the case. Judgments of the board will be final and binding on the parties to the dispute. They will be "enforceable *per se* in the territories of the members."[20]

Certain other types of cases can also come before the OAPEC Judical Board provided that both sides in the dispute agree that the board should hear their case. Such cases include disputes between a member and a petroleum company, either a privately owned company or one of the national oil companies of OAPEC members; and disputes between members that do not fall in the field of petroleum. Judgments in these cases will be equally binding. In effect, the board will be empowered to operate as an international court of law in cases where recourse to a member's national courts might be regarded as intrinsically prejudicial to one side.

The protocol authorizing the establishment of the OAPEC Judicial Board was passed by the council at its December 1977 meeting. *The Fifth Annual Report* of OAPEC's Secretary General, published one year later, reported that eight OAPEC members had ratified the protocol.[21] At the June 1980 meeting of the council, the protocol was reported as having been ratified, and the ministers announced that they would appoint the first judges at their next meeting.[22] For the most part, there has been little said about the progress toward the ratification of the protocol. I have been unable to find out either the order of member ratification, or reasons why ratification took two and a half years. However, one obvious reason is that an international agency with power to make and order enforced redistributive decisions would directly threaten OAPEC member sovereignty. For a country such as Saudi Arabia, the desire to preserve its own national power would compete strongly with the desire to support a united Arab front in petroleum affairs.

Neither OPEC nor the Arab League has charter provisions authorizing the establishment of judicial bodies. The OPEC Conference has discussed the possibility of having an OPEC court several times, but no action implementing such a policy has ever been taken. The problem of designing a court compatible with OPEC is similar to the problem of production prorationing. Decisions in both areas create OPEC winners at the expense of OPEC losers. If OPEC can begin to permit decisions that are not unanimous or if it can broaden its mandate so that packages of agreement can be forged among its members in such a way that the same members can win and lose within a single package, it may eventually be able to make decisions involving redistribution of benefits among its members. Until then, such issues will continue to lie beyond the ability of OPEC to act upon.

These institutional differences reflect differing self-images and goals among the three organizations. The Arab League is committed to preserving member sovereignty. It is primarily concerned with intra-Arab relations and Arab policy toward outside nations. The Arab League decision rule is that member sovereignty is to be preserved above all else. The Arab League does not see itself as governmental but rather contents itself with the limited role of fostering cooperation among its members. A court would be incompatible with league goals.

OPEC is an economic organization of oil producers that operates as a cartel. Its members' aim is to work together to obtain the highest possible return for their resources. Success for a cartel can be measured by its ability to increase and maintain prices significantly above marginal cost,[23] something OPEC has done increasingly well since Libya's 1970 victory in negotiations with its oil companies. The greatest potential for division within OPEC is over the problem of allocating production shares in situations where significant excess capacity exists. Until now, production limitation has been ad hoc and voluntary. Voluntary restraint is achievable because oil prices are high enough so that OPEC members can keep their incomes at acceptable levels while producing below capacity.[24] In some sense production restraint within OPEC operates on the basis of tacit collusion among members.[25] Historically, other cases of producer collusion have resulted in cartel collapse after an average period of from six to eight years owing to the accumulation of dissatisfaction on the part of individual members and the inability of the cartels to enforce their market-sharing provisions.[26] Tacit collusion in OPEC has prolonged OPEC's success because it enables members to ignore cheating, at least for a while. But the problem of dividing finite benefits among its members will unavoidably arise if OPEC is no longer able to supply benefits to members at the expense of outsiders (see Chapter 5).

OAPEC sees itself as "in fact, an Arab common Market . . . with all the commitments and rules it involves."[27] It can own property, sign contracts, and sue and be sued.[28] It thus has a corporate identity closer in nature to that of the European Common Market than to that of OPEC. OAPEC avoids conflict with OPEC by making all OPEC resolutions automatically binding on OAPEC members, whether or not they are also members of OPEC.[29] Perhaps as important, by conceding primacy to OPEC with regard to pricing and production policy making, OAPEC avoids conflict within itself. Instead it concentrates on how best to invest some of the proceeds of OPEC's success. These tasks are inherently more pleasant than those required for cartel management and less divisive because the benefits OAPEC generates accrue to all its members, although not equally. Despite inequality, the allocation of benefits and costs of OAPEC projects appears to be regarded by its members as equitable most of the time (see Chapter 4).

The Implementation of OAPEC Policies

Article 2 of the OAPEC Agreement outlines the ways in which the organization intended to achieve its objectives of promoting member cooperation

in various forms of economic activity in the petroleum industry, the realization of the closest ties among them in this field, the determination of ways and means of safeguarding the legitimate interests of its members in this industry, individually and collectively, the unification of efforts to ensure the flow of petroleum to its consumption markets on equitable and reasonable terms, and the creation of suitable climate for the capital and expertise invested in the petroleum industry in the member countries.

In pursuit of the said objectives, the Organization shall in particular:

(a) Take adequate measures for the coordination of the petroleum economic policies of its members.

(b) Take adequate measures for the harmonization of the legal systems in force in the member countries to carry out its activity.

(c) Assist members to exchange information and expertise and provide training and employment opportunities for citizens of member countries in members' countries where such possibilities exist.

(d) Promote cooperation among members in working out solutions to problems facing them in the petroleum industry.

(e) Utilize member resources and common potentialities in establishing joint projects in various phases of the petroleum industry such as may be undertaken by all the members or those of them that may be interested in such projects.[30]

These objectives and methods can be divided into two groups. The first group consists of provisions a, c, and d. These are "traditional" aims and objectives similar to those of the Arab League or OPEC. They are passive objectives and methods, requiring OAPEC to "coordinate," not to make, oil policy; to "assist" members in exchanging information and providing training and employment, not to assume the central responsibility and control over these functions; and to "promote" cooperation rather than to coerce compliance. These provisions are permissive in that they leave initiative with individual members and reflect the tension between member autonomy and centralized power.

The other two provisions, b and e, are novel inclusions. The mandate they give to OAPEC requires the organization to be active: to "take active measures" to coordinate member legal systems rather than merely to promote such measures and to "utilize member resources" to establish joint ventures rather than simply to encourage such a use. In response to this mandate OAPEC has created novel institutions to implement the charter provisions. These institutions are the joint venture companies and the Judicial Board.

Passive Implementation

OAPEC's role as coordinator of member petroleum policies was designed in part to enable Arab oil producers to cope with two kinds of pressure on them. One was the demand by other Arab League members to use oil as a political weapon against Israel. The use of the oil weapon in 1967 resulted in significant losses of income to the oil-producing Arab states. The OAPEC founders hoped that their organization would impress upon other Arab states the need to consider the economic as well as the political consequences of manipulating Arab oil production. OAPEC also served as another institutional means of encouraging oil-producing country solidarity as a way to combat the tendency of oil-exporting countries to permit their oil to be used to exploit other oil exporters. During conflicts between an oil-exporting country and the oil companies operating on its territory, the companies usually decreased their take of oil from that country as a means of exerting pressure through denial of income. The companies did not suffer themselves because they could make up the shortfall by increasing production somewhere else.[31] Before oil prices increased dramatically in the early 1970s, selling extra oil was very tempting to low-income oil-producing states, even though doing so permitted the companies to play countries off against one another. OAPEC was seen as another forum in which member petroleum policies might be aligned and producer solidarity strengthened in the face of attempts by the companies to divide the oil-exporting countries.

At first OAPEC took an active role in member oil-policy coordination. Subsequent to the first OAPEC Council meeting, a single OAPEC delegation attended the October 1968 meeting of the Arab Oil Experts Committee, an agency of the Arab League. According to Saudi Oil Minister Ahmad Zaki Yamani, " . . . [This is] in accordance with [a] decision of [OAPEC's] Council that henceforth the three countries will maintain joint representation at Arab League petroleum meetings, and that future communication between Arab League oil agencies and the three countries in this regard will be channelled through OAPEC."[32] This underscored the position of the OAPEC founders regarding Arab League interference in oil production policy. They would not just present a united front but would actually send a single delegation to Arab League oil meetings.

At the next Arab Oil Experts Committee meeting in January 1969, Egypt proposed the establishment of an Arab petroleum organization under the aegis of the Arab League, clearly a response to the OAPEC unitary delegation. The same meeting produced proposals by individual non-OAPEC delegations to create project companies similar in structure and identical in purpose to those currently under consideration by OAPEC. One of these proposals was to create a joint Arab tanker company.[33] OAPEC's response was to remain aloof. An unnamed source told the *Middle East Economic Survey* that while OAPEC members would welcome the creation of an Arab petroleum organization they would not join it because they already had an organization of their own. *MEES* also learned that OAPEC was not "prepared to associate itself with projects that were similar to plans it was contemplating itself."[34] The lines were drawn in preparation for a battle over which organization, OAPEC or the Arab League, was to control Arab oil policy.

As it turned out, the battle was fought almost entirely within OAPEC itself. The 1 September 1969 coup in Libya produced a revolutionary regime that was extremely interested in pursuing a vigorous oil policy.[35] The other two members of OAPEC were initially conciliatory toward the new regime, which came under the leadership of Mu'ammar al-Qadhafi on 16 January 1970. The fourth OAPEC Council, meeting on 26-27 January 1970, appointed Libyan Suhail Sa'dawi its first secretary general.[36] It was later revealed that the council had also resolved:

That each of the Organization's member states shall be represented individually at oil meetings of the Arab League, on the understanding that there shall be prior coordination between the positions of the Organization's member states regarding important oil matters which may be under consideration in both OAPEC and the Arab League Secretariat General, so as to enable the Secretary General of OAPEC to attend Arab League oil meetings in an observer's capacity (Resolution IV. 10).[37]

Thus OAPEC's united front policy within the Arab League was implemented for only a short time. We know that for a resolution to pass the council, a three-fourths majority must vote in its favor. When three is the total, three is also the number needed to meet a three-fourths-majority voting requirement. Whether Libya, Saudi Arabia, or Kuwait proposed the resolution, both moderate members could see its value since the government and the interests of Libya had changed. But this action effectively reduced OAPEC's ability to take independent positions on oil policy issues within either the Arab League or OPEC. By the time OAPEC recovered from its membership crisis, also stimulated by the changes in Libya's government, OPEC had begun a dramatic series of confrontations with the oil companies. Oil producer policymaking with regard to oil prices and production remained in the OPEC arena despite some subsequent discussion of oil pricing within OAPEC. Because of intra-Arab disagreements over pricing policy, OAPEC's implementation of both the spirit and the letter of the provision to defer to OPEC's price decisions probably contributed significantly to solidarity within the Arab organization by removing a deeply divisive issue to another arena.

OAPEC now implements provision a of Article 2, the oil policy coordination provision, by contributing organizational support to member countries in conflict with the oil companies[38] and by sponsoring feasibility studies of possible regional downstream investments. Recent examples are OAPEC's research into the desirability and feasibility of establishing Arab industries to manufacture lubricating oils and synthetic rubber to meet the regional demand for these products.[39] OAPEC has taken strong public positions encouraging member downstream investment, particularly in refining, but these exhortations have been general and have not dealt with the problems of coordinating such investment. Thus OAPEC plays an ambiguous role in this type of situation. On the one hand, the organization appears to be trying to unify its members by construing all producer efforts to invest downstream as blows against imperialistic exploitation by multinational corporations based in developed countries.[40] On the other hand, the editorials rarely enjoin members to a collective, coordinated downstream investment policy, their silence a form of consent to the recent pattern of nationalistic investment.[41]

Provision c of Article 2 of the OAPEC Agreement authorizes OAPEC to assist in training Arab nationals for positions in the region's petroleum industry and to facilitate information exchange among the member countries. The training of Arab nationals has been a major concern of OAPEC since its foundation. OAPEC joint projects, such as the Arab Shipbuilding and Repair Yard and the Arab Maritime Petroleum Transport Company, were designed to be able to train technicians for their own companies and eventually for employment in other parts of the petroleum industry. In November 1975, the OAPEC Council recommended that the secretariat establish a division for manpower training and development to organize training courses and seminars.[42] The secretariat did a detailed study of manpower in the oil and gas industry in member countries and followed its report with a series of forecasts of future manpower needs in the region's oil sector. The creation of an OAPEC training institute has been approved by the council as a means to help meet the need for technical skills in the working population.[43] OAPEC also sponsors training seminars designed to introduce new entrants into the oil industry to "basic aspects of the international oil and gas industry in general and the Arab industry in particular," and it offers specialized courses on technical elements of the oil industry.[44]

OAPEC facilitates the exchange of information about the Arab petroleum industry in a variety of ways. The secretariat prepares an annual statistical report on the petroleum industry in member countries. General information is gathered through a library and documentation service that supervises a collection of books, periodicals, and research reports for use by the OAPEC staff, member countries, and visiting scholars. The library cooperates with other libraries, with the petroleum ministries in member countries, and with Arab and other international organizations in information exchange programs.

OAPEC has published a monthly news bulletin since November 1975, and a journal, *Oil and Arab Cooperation*, which has been issued quarterly since 1975. The secretary general has published an annual report since 1975. The report, which is available in Arabic, English, and French, includes information on one or more themes related to the oil industry generally and a detailed report of the activities of the organization for that year.

OAPEC staff members attend international conferences and seminars on oil, economic development, and international cooperation, many of which OAPEC sponsors or cosponsors with other international organizations, universities, and business groups, including those from the industrialized West.[45] OAPEC staff members often present papers dealing with the organization itself and the way it sees its role as an agent of development and cooperation in the region. OAPEC also sponsors specialized seminars where scholars and industry personnel are invited to present papers on a specific aspect of the industry as a sort of fact-finding exercise for OAPEC in its quest for additional areas in which to mount joint ventures.

Provision d of Article 2 calls for OAPEC to promote cooperation among its members in seeking solutions to petroleum industry problems. There is some indication that OAPEC is interested in becoming the coordinating agent of the national oil companies. Such a role would give OAPEC the potential to coordinate member exploration, production, and sales. As the oil-exporting countries assume greater control over their oil, they will need to coordinate their production to avoid price competition and their exploration to avoid future competition and income inequities. Under the pre-1973 world oil regime, these functions were performed by the foreign oil companies. Exploration and production decisions were made according to a process whereby the operating companies (such as Aramco) submitted proposals for exploration and production to their parent companies (such as Exxon or Texaco), which made their final decisions based on estimates of demand and the field reports of the operating companies. The operating companies acted as buffers between the oil-exporting countries, each of which wanted to increase its own production for income purposes, and the multinational oil companies, which wanted to pump enough oil to meet demand but not so much that prices would be depressed. This system was similar to the one made up of the entire complex of oil companies acting as a buffer between oil exporters and oil consumers. A major function of both of these buffer systems was to reduce opportunities for conflict between contending parties by introducing a third party that shared some of the goals of each principal. This gave the principals someone else to blame when they acted in ways that might have invited retaliation by the other side.

At least one member of the OAPEC Secretariat has expressed concern that the "new petroleum order" institute a buffer system to "depoliticize" exploration and production decisions when the national oil companies take full control of operations from the present operating companies.[46] In one scenario of the new order, the national oil companies are cast in the role of operating companies for the oil-producing country governments, which play the part of the multinationals and make oil policy decisions. Presumably the national oil companies could introduce an important economic and technically oriented dimension into oil policy decisions, which the governments have tended in the past to treat in a political context. However, at this time, the national oil companies have no institutional framework for coordinating their individual decisions on production. Under the old system the international parent oil companies could share information and coordinate production because their interests in various operating companies overlapped.[47] The national oil companies do not have a cross-national structure because each operates only in a single country. Arab joint ventures have tended to be limited in scope and are not vertically integrated. Their ranges of operation are also limited. The Arab joint venture projects by themselves are not adequate to the task of institutionalizing an exchange of information sufficient to coordinate the production and exploration policies of the oil-exporting countries.

Former OAPEC Assistant Secretary General Mahmoud Amin made two proposals for using OAPEC as the coordinating instrument, either as an agent to coordinate the "petroleum economic policies in the Member states through the petroleum ministries" or as the technical arm of a separate organization of the national oil companies.[48] It is, of course, the petroleum ministers of the OAPEC member states who sit on its council, on the OPEC Conference, and on the various oil policy committees and agencies of the Arab League. These individuals are vulnerable to cross-pressures by the very fact of their multiple memberships in organizations whose policies occasionally diverge.

The most likely source of tension for an oil minister, however, is his simultaneous obligation to his government and to the international oil organizations. Because oil is so critical to the economies of most OAPEC countries, the oil minister is a central figure in his home government. He is not just any technocrat: he supervises the economic sector that supports everything his government does.[49] The type of person the neofunctionalists would have assumed to be the most likely agent of "spillover"[50] increasing the scope of international integration turns out, in this instance, to be the policymaker facing the heaviest domestic demands.

The inability of the oil ministers to increase international integration by expanding the scope or the range of the international oil organizations has been shown within OPEC. OPEC has been unable to agree upon a plan for production prorationing even though pricing policy and thus total production is part of its organizational mandate.[51] This is because the oil ministers cannot reconcile domestic desires for money, autonomy, and prestige with OPEC's desire to be assured of an aggregate level of production low enough to support its current price policy. Production prorationing becomes an issue in OPEC whenever other factors do not intervene to tighten oil supplies. For example, prior to the 1979 apparent shortfall in world oil production caused by the crisis in Iran, OPEC set up yet another committee to consider the prorationing issue. The disintegration of OPEC pricing policy since then greatly politicized the prorationing issue, and an agreement on prorationing still appears far from achievement.

A similar situation governs the behavior of the oil ministers in OAPEC. They have been fairly generous in locating projects throughout the region, but this represents a minimal degree of redistribution in the context of Arab nationalism and Islamic ethics[52] and their concrete implementation through richly endowed Arab and Islamic development funds. Asking the oil ministers to coordinate petroleum economic policy for their members would be forcing them to face more fundamental redistributive questions, even if they had only to suggest rather than to mandate production targets in the member states.

Somewhere between project location and oil production allocation on a scale of redistributive policies lies member support of functioning joint venture projects. OAPEC has not yet been able to muster more than half-hearted member support for its joint ventures once they have been launched. It could not prevent the establishment of national tanker fleets in member states in the face of plans for the OAPEC joint venture tanker company. The OAPEC Council of Oil Ministers waited until September 1978 to protect the joint tanker company through member subsidies and guarantees of business.[53] Action to protect an investment already in place has to precede any credible attempt on the part of OAPEC to perform production coordination. This process appears to have begun with the candid assessment of member support of OAPEC by Secretary General Ali Attiga at an OPEC seminar in October 1977.[54] A big step was taken in September 1978, when the Council of Ministers voted to require members to support the Arab joint tanker company. Smaller, yet important, steps have involved the formation of two committees charged with assessing the need for and feasibility of expanding OAPEC's role as oil policy coordinator. They are the Standing Committee on Refining, which has sponsored a study of proposals for a joint Arab refining policy, and the Legal Committee on Maritime Transport, which investigates the possibility of establishing an Arab legal regime to govern maritime transport of hydrocarbons. The Committee on Refining aims to coordinate Arab development in the refining industry by evaluating demand forecasts for the region and facilitating the exchange of information on products supplied by the various refineries in Arab countries. Its aim is to avoid inter-Arab competition in refined products and to establish a home market base for the Arab refining industry. The Legal Committee on Maritime Transport was formed before the Council of Ministers voted to support the Arab joint tanker company, and during its early months was concerned with establishing rules for all Arab oil-exporting countries mandating preferences for Arab tanker fleets, including the fleet of the OAPEC joint venture company. Since September 1978, few details of the meetings of this committee have been published.

Finally, a special unit to coordinate exploration and production was set up by the OAPEC Secretariat in early 1979. The new unit has wide-ranging functions extending from data collection to the development of standards for evaluating member efforts in oil exploration. The unit will have a strong research arm involved in technical studies, and like most other OAPEC agencies, it will be responsible for conducting training courses and seminars.[55] If the unit can establish itself as an authority in drilling and exploration, and can provide related services to meet member demand, it may have the opportunity to expand OAPEC's coordinating role.

We can see that in the important area of coordinating member oil policy, OAPEC has only begun to function in an effective manner. OAPEC has been more successful in its aims to train an indigenous work force at all levels of the Arab oil industry. It has also been instrumental in facilitating the flow of information among its member countries, a role it shares with OPEC and, to some extent, with the Arab League. Part of the problem in achieving organizational objectives in these instances is the passive role assigned to OAPEC as reflected in the provisions of Article 2 of the OAPEC Agreement. OAPEC can only offer its services in education, information exchange, and oil policy coordination. It has no coercive power to make its members adopt a collective approach. A somewhat different environment surrounds OAPEC implementation of the provisions of Article 2 involving the joint venture projects and the harmonization of member legal systems.

Active Implementation

Several provisions of the OAPEC Agreement deal with the establishment of joint venture companies by the organization.[56] From 1972 through 1977, four companies involved with some aspect of the oil industry were launched.[57] These are listed in Table 3. Since then, a fifth company and two subsidiaries of the Arab Petroleum Services Company have been established. Each was created according to the description in Article 2-e of the OAPEC Agreement permitting the organization to establish "joint projects in various facets of the petroleum industry which may be undertaken by all the members, or by those of them that may be interested in such projects."[58] The companies were set up under the direction of the OAPEC Council of Minis-

TABLE 3 The OAPEC Joint Investment Projects, 1972-1977

	AMPTC[a]		ASRY[b]		APICORP[c]		APSC[d]	
	% of Shares	Value $ mil	% of Shares	Value $ mil	% of Shares	Value $ mil	% of Shares	Value $ mil
Saudi Arabia	13.56	67.8	18.84	64.0	17.00	57.8	14.00	7.1
Kuwait	13.56	67.8	18.84	64.0	17.00	57.8	14.00	7.1
UAE	13.56	67.8	18.84	64.0	17.00	57.8	14.00	7.1
Iraq	13.56	67.8	4.70	16.0	10.00	34.0	3.00	1.5
Qatar	13.56	67.8	18.84	64.0	10.00	34.0	10.00	5.1
Bahrain	5.00	25.0	18.84	64.0	3.00	10.2	3.00	1.5
Libya	13.56	67.8	1.10	3.8	15.00	51.0	17.00	8.6
Algeria	13.56	67.8	—	—	5.00	17.0	10.00	5.1
Egypt	0.10	0.5	—	—	3.00	10.2	5.00	2.5
Syria	—	—	—	—	3.00	10.2	10.00	5.1
Authorized Capital*	500,000,000		340,000,000		1,019,830,029		337,780,000	
Subscribed Capital*	500,000,000		340,000,000		399,943,343		50,667,000	
Paid Capital*	411,299,841		270,485,714		169,971,671		25,333,500	

Source: Ali A. Attiga, "Regional Cooperation in Downstream Investments: The Case of OAPEC" (paper delivered in the OPEC Seminar on the Present and Future Role of the National Oil Companies, Vienna, 10-12 October 1977), Table II.

[a] AMPTC: Arab Maritime Petroleum Transport Company, founded 1972.
[b] ASRY: Arab Shipbuilding and Repair Yard Company, founded 1974.
[c] APICORP: Arab Petroleum Investments Corporation, founded 1975.
[d] APSC: Arab Petroleum Services Corporation, founded 1977.
*Value in U.S. Dollars.

ters. The council determines the total authorized (permitted) capitalization for each country and supervises the distribution of shares among the members in each one.[59] Once a company charter has been drawn up by representatives of the participating countries, the Council of Ministers becomes its General Assembly.[60] The General Assembly announces the establishment of the company, selects its first board of directors and invites participating member states to pay their shares of the capitalization.[61] Then each company becomes independent of OAPEC in a legal sense, although the council as the General Assembly, composed only of ministers from its shareholding states, supervises its activities. The legal independence of the companies gives them a bit of distance from the politics of the shareholding governments. They are not isolated but rather liberated from disagreements among OAPEC member governments.

The joint venture companies are novel institutions. Although they are set up to have something of a "private" economic character,[62] their charter instruments are basically international agreements among the OAPEC member states which are their shareholders. There is a voluntary aspect to member participation in three of the joint venture companies: member governments may participate or not according to their individual circumstances and desires.[63] The separation of the management of each company from the shareholding governments, plus provisions in the company charters giving the legal system outlined in the charter precedence over any national law in any member state, puts the shareholding governments in an ambiguous position relative to the joint venture companies. The shareholding governments are suppliers of capital and custodians of their alloted shares in the companies.[64] They collect any dividends declared by the companies. But they may not control the behavior of the companies, even if the companies operate within their territorial jurisdictions or declare themselves "nationals" of one or more of the participating states.[65] In addition, the shareholding governments have positive responsibilities toward the companies, the most important of which is their obligation to exempt the companies from taxes, duties, and fees regardless of whether such payments are required from other companies operating in the territories of the states.[66] Each of these restraints on and obligations of the shareholding governments in their relations with the OAPEC joint venture companies is part of the companies' charters and has the force of any other international agreement among the participating states. The OAPEC joint venture companies thus enjoy a greater degree of institutional protection from the governments than other companies operating in the region. Their obligations to the shareholding governments, aside from fulfilling the specific aims set forth in their charters,[67] are mediated by the OAPEC Council of Ministers. Each company must submit its annual report to the council and must operate under its general policies and directives. Beyond that, the companies are independent entities.

The joint tanker company, the Arab Maritime Petroleum Transport Company (AMPTC), was the first of the OAPEC joint venture projects. None of the Arab oil-exporting countries had a tanker fleet when AMPTC was chartered.[68] Since then, OAPEC members have started national tanker fleets that compete directly with the joint venture company. AMPTC had a capacity of 2,086,000 dead weight tons (dwt) in its fleet of eight tankers according to the company's 1977 annual report. During that year it had 105 employees and maintained its policy of increasing the proportion of Arab workers in technical positions. A total of 216 trainees sponsored by AMPTC were sent to maritime academies in three Arab countries and the United Kingdom.

The Arab Shipbuilding and Repair Yard Company (ASRY) opened its dry dock complex in Bahrain on 23 October 1977. This dry dock is the only such facility between Portugal and Singapore capable of servicing very large crude carriers (VLCCs).[69] Its strategic location in the Arabian Gulf at the confluence of VLCC traffic where ships may put up for repairs in a gas-free condition (the ships, in effect, are vented on their return trip to the Gulf to take on more oil) and its competitive pricing policy have guaranteed its operation near or at peak capacity despite periods of depressed market conditions in ship repairing. ASRY has had to cut prices below actual costs and sustain operating losses to keep its facilities running and its workers employed. But ASRY Chairman Sheikh Khalifah bin Salman bin Muhammad regarded his instructions from OAPEC as stressing the strategic nature of the dry dock rather than profits in its early years. "At the present time, there are operating losses. They were foreseen since 1973. . . . But the project was still considered feasible in terms of creating an industry, and training Bahrainis and other Arab nationals, in a downstream operation of the oil industry. It was also considered as a strategic project, in the sense of meeting the needs of OAPEC's joint fleet and the fleets of member states."[70]

The first Arab ship docked at ASRY for repairs in January 1978, and in March of that year an AMPTC vessel used the yard for the first time.[71] By the end of 1978 the shipyard was operating at 94 percent of capacity. ASRY supported 400 trainees in technical and management fields and employed a total of 1200 persons, 43 percent of whom were reported to be Arab nationals.[72] Among the facilities at the shipyard are several steel workshops, which ASRY's management is interested in expanding to supply steel to the local market and eventually to the entire region.[73]

The Arab Petroleum Investments Corporation (APICORP) was founded in November 1975 as an alternative to OAPEC's becoming directly involved in oil refining and petrochemicals manufacturing at that time. APICORP supplies capital in the form of equity investment and loans to downstream activities in Arab countries. However, APICORP found itself shut out of some investment opportunities "due to the highly exclusive policies followed by its shareholders [the OAPEC member states] with regard to foreign investment in their petroleum industries."[74] The national oil companies prefer taking loans from APICORP rather than allowing the OAPEC company to enjoy equity participation, an attitude not unlike that of their predecessors, the operating companies, with regard to the oil-exporting country governments. In spite of its problems in finding projects willing to accept its investment assistance and its difficulty in recruiting personnel to staff its operation,[75] APICORP participated in a number of projects in 1978: a fertilizer project in Jordan; a natural gas project in Bahrain; a joint project with the Arab Petroleum Services Company (APSC), another OAPEC joint venture company concerned with oil drilling; a detergent manufacturing project; a catalysts manufacturing project; and a lube oil project. APICORP initiated the work in the last three project areas mentioned.[76]

The fourth OAPEC joint venture company is APSC. Its charter went into effect in August 1976 and its first board of directors was named the following January. APSC is strictly a holding company, as opposed to the other three companies whose charters permit them to function as both operating and holding companies. APSC was designed to control a group of companies, each with a distinct corporate personality and each concerned with some aspect of petroleum services. The APSC subsidiaries need not be wholly owned. The APSC charter stipulates that the company and one or more national companies of its member states must hold a minimum of 60 percent of the equity of any subsidiary. Foreign partners may also participate in APSC subsidiaries. APSC established its first subsidiary in October 1977, the Arab Drilling and Workover Company (ADWC),[77] currently engaged in drilling operations in Libya. ADWC has already begun its own training program.[78] A well-logging company was set up in 1980.

In spite of its heavy emphasis on personnel training, OAPEC did not establish an agency specifically for training until 1978. On 9 May 1978, the OAPEC Council of Ministers agreed to set up the Arab Petroleum Training Institute (APTI) to be headquartered in Baghdad. According to *MEES*, a conservative estimate of the capital cost of this project is $12 million.[79] The institute offers courses for graduates of training institutions in their own countries, preferably those who have had some experience working in the oil industry.[80] APTI graduates then return to their home countries and train the new workers who are needed if the oil industry in Arab countries is to be operated by indigenous personnel.

The OAPEC Judicial Board is not, like the joint venture companies and the training institute, an OAPEC "project." Provisions were made for such an institution in the original OAPEC Agreement. Yet the OAPEC Council of Ministers waited for ten years before they signed a protocol detailing the manner of its establishment and operation and sending it to the member states for ratification. Implementation of the protocol marks a developmental milestone in the life of OAPEC. Because the Judicial Board will have compulsory jurisdiction in disputes involving member governments or OAPEC joint venture companies in all cases pertaining to OAPEC's activities, the approval of the members with regard to its establishment signifies a willingness on their part to surrender some of their sovereignty in favor of an agency of OAPEC.

Although the details of the operation of the Judicial Board are not yet available, the broad outlines of its mandate are contained in the OAPEC Agreement. In addition, each of the charters of the OAPEC joint venture companies contains a provision granting it jurisdiction over disputes arising between the member states with regard to its application, interpretation, or execution. The Judicial Board will fit into a carefully prepared niche within the structure of OAPEC and its dependent and independent agencies and organizations. Ratification of the protocol was the final step in the process of its evolution from idea to institution.

OAPEC and Economic Development

The Arabs in general, it seems to me, have always had an ambivalent view of the West. They want our technology and practical "secrets"; but they do not want to lose their own culture and social strengths. OAPEC is just one of the devices they have used to cope with the oil problem since the takeover of the Western oil interests . . .

(Georgiana Stevens, *letter to the author,*
9 May 1978)

The OAPEC joint venture projects concentrate on economic development in the petroleum industry. AMPTC and ASRY are pioneers in Arab downstream investment. ASRY is the most successful of the OAPEC projects. It is very close to being a profitable operation and has no shortage of business. Because of the Bahrain installation's success in operating near or at capacity since it opened, OAPEC has agreed to the construction of a second dry dock complex in one of its Mediterranean member countries, Algeria.

The ASRY dry dock in Bahrain was constructed specifically to serve the enormous tankers that carry crude oil all over the world. Eventually, when oil depletion makes the monster ships obsolete, the dry dock can be used for smaller vessels. ASRY has trained hundreds of Arab nationals (see Table 4). Its small steel plant is a venture into nonpetroleum industrialization; it is now capable of supplying steel to operations outside the dry dock itself.

AMPTC, on the other hand, is closely tied to the oil industry. Its workers may be able to transfer their skills to other kinds of ships, but AMPTC's ever-increasing capital investment is solely in oil and gas carriers.[81] AMPTC has been relatively unsuccessful in its efforts to operate at or near capacity. This is owing as much to investment duplication by the member governments as to the unwillingness of consumers to transport their oil in Arab tankers. Duplication more than concentration in an industry with a limited life span

TABLE 4 *A Partial Listing of OAPEC Trainees, 1974-1979*

Date	Number	Institution
1974	93	AMPTC
1975	1	SECRETARIAT
	55	AMPTC
1976	157	ASRY
	206	AMPTC
1977	246	ASRY
1978	149	ASRY
	220	AMPTC
	14	SECRETARIAT
1979	180	ASRY
	12	AMPTC

Source: OAPEC *Bulletin,* various issues; OAPEC *Annual Report of the Secretary General,* various issues. Data for AMPTC and OAPEC Secretariat probably incomplete.

retards development in the Arab world. Arab tanker fleets have proliferated since OAPEC's decision to create the first Arab tanker company. In 1978, Kuwait, the OAPEC member state where AMPTC's headquarters is located, had twelve tankers of its own in addition to the AMPTC oil tanker and gas carrier which also flies its flag; Saudi Arabia, home of the last two board chairmen of AMPTC, had seventeen other tankers in national fleets.[82]

In spite of a weak tanker market, OPEC members and outsiders have continued to add to their tanker capacity. From 1975 to 1978, a period of excess tanker capacity at 1975 fleet levels, non-OPEC sources, including oil companies and private owners, "added to their tanker fleets capacity amounting to three times that of the entire OPEC tanker fleet."[83] This expansion foreshadows an eventual showdown between OPEC members and non-OPEC operators of tanker fleets for shares of the maritime transport industry. Regardless of the outcome, transport profits are likely to remain small or nonexistent. This means that AMPTC and the Arab national tanker fleets will continue to drain limited OPEC-OAPEC member resources, instead of generating profits for the governments to use for other development projects.

The Arab Petroleum Investments Corporation has generated profits for its shareholders.[84] APICORP's contribution to economic development in the Arab Middle East extends beyond the petroleum industry and beyond the borders of the OAPEC member states. APICORP channels funding for commercial projects from capital rich to capital poor Arab countries. This spreads OAPEC's wealth to the resource poor and to the less wealthy among themselves such as Bahrain, whose oil is on the verge of depletion.

APICORP is interested in profits. Its capitalization of industrial projects complements conventional foreign aid, much of which goes to infrastructural projects, public works necessary to a developing economy. APICORP enables Arab governments and corporations that accept its loans and equity participation to avoid some of the crippling byproducts of economic dependency.[85] It is a local alternative to direct foreign investment by developed countries or their multinational corporations. Financing a project through APICORP saves foreign exchange that would leave the project site country if outsiders repatriated their profits. Financing through APICORP also keeps control of projects in the region.

The Arab Petroleum Services Company is another petroleum-related company, but its subsidiaries concentrate specifically on lessening dependency on the multinational oil companies rather than depending on foreign sources of investment capital. ADWC does exploratory and developmental drilling and oil well maintenance work. These are "nuts and bolts" aspects of a functioning oil industry. So is electric well-logging, which will be done by APSC's second subsidiary.[86] APSC is investigating other, similar activities for future subsidiaries.[87]

Dependence on multinational oil companies is more than an economic relationship; it creates political dependency as well on the multinational's home country. As the Iranian experience in 1980 after the imposition of the United States economic sanctions has shown, local industries may be crippled by the results of government-to-government conflicts as easily as by government-to-company conflicts. If such sanctions were to be instituted against OAPEC countries, the Arab oil industry would not be able to perform all required maintenance nor to replace all needed spare parts. APSC has been set up to reduce and eventually to eliminate this kind of dependency.

Technology Transfer through OAPEC

It is still too early to say that technology transfer has occurred in the Arab world, much less to measure it and then decide who or what was responsible. One indicator of self-sufficient technology in a country or region is scientific activity. While Arab scientific activity has increased since the oil revolution of 1973, it is impossible to determine the amount of such activity directly attributable to OAPEC. Easier to count although less indicative of technology transfer or development are specific OAPEC activities directed toward the goal of technology independence. Two such activities are training programs and professional conferences and seminars.

OAPEC trainees have been instructed by every joint venture project company. OAPEC's newest training facility, APTI, is devoted entirely to the education of middle- and upper-level personnel in the petroleum industry. It began its first training programs in 1980. APTI was specifically intended by the organization to act as an agent of technology transfer and development.

OAPEC's commitment to training at all levels has survived operating losses by AMPTC and ASRY and political conflict over the siting of the Petroleum Training Institute (see Chapter 4).

Conferences and seminars are another important OAPEC contribution to technology transfer. Professional meetings are opportunities for scientists, technicians, and managers to exchange views and information. OAPEC seminars have contributed to professional development and also to advancing the frontiers of the Arab oil industry by serving as occasions for research into the feasibility of various downstream investments. Table 5 lists the seminars sponsored by OAPEC, alone and jointly with one or more other organizations.

TABLE 5 *OAPEC-Sponsored Seminars and Conferences, 1974-1980*

Date	Title and Place
1974	
May	Opportunities for Cooperation Between Great Britain and the Arab World* (London)
1975	
January	Training Seminar on Liquified Gas Technology and Trade (Kuwait)
October	Seminar on Prospects of Arab Refining Industry* (Damascus)
November	Opportunities for Cooperation Between France and the Arab World* (Paris)
1976	
January-March	Training Seminar on Fundamentals of the Oil and Gas Industry (Kuwait)
October-November	Seminar on Petrochemicals* (Alexandria)
November	Opportunities for Cooperation Between Japan and the Arab World* (Tokyo)
1977	
May	Seminar on Arab Oil Information (Kuwait)
November	Seminar on Arab Lube Oil Industry (Alexandria)
December	Seminar on Reservoir Engineering (Kuwait)
1978	
January-February	Training Seminar on Fundamentals of the Oil and Gas Industry (Kuwait)
June	Seminar for Managers and Experts of Training Centers in OAPEC Member States (Algiers)
October	Seminar for Arab Refining Managers (Kuwait)

Table 5 continued

Date	Title and Place
1978	
November	Seminar on Petroleum Exploration (Kuwait)
December	Seminar on Development Through Cooperation Between OAPEC and the Scandanavian Countries* (Oslo)
1979	
January	Seminar on Petroproteins (Kuwait)
March	First Arab Energy Conference* (Abu Dhabi)
April-May	Training Seminar on Fundamentals of the Oil and Gas Industry (Kuwait)
May-June	OAPEC Reservoir Engineering Seminars (Algiers)
September	First Oxford Energy Seminar* (Oxford)
1980	
February-March	Training Seminar on Fundamentals of the Oil and Gas Industry (Kuwait)
March	APTI Seminar for Directors of Arab Training Centers (Baghdad)
June	APTI Conference on Inspection and Industrial Safety (Baghdad)
June-July	Seminar on Ideal Uses of Natural Gas in Arab Countries* (Algeria)
September	Survey of Refineries in Syria and Jordan
	Seminar on Benzene Reforming and Catalytic Hydro-treatment and Hydrocracking (Damascus)
November	Seminar on Maritime Transport (Kuwait)
	Seminar on Reservoir Engineering (Kuwait)
Unscheduled	Second Oxford Energy Seminar*

Source: *OAPEC Bulletin*, various issues, and *Annual Report of the Secretary General*, various issues.

*Cosponsored.

The OAPEC-sponsored seminars and conferences are notable for their subject matter development. An example is the seminar on fundamentals of the oil and gas industry, held yearly since 1978. This is the most highly institutionalized seminar sponsored by OAPEC. Participants are middle-level managers of national oil companies and member government oil ministries. The first seminar on fundamentals of the oil and gas industry (Kuwait, 1976) was taught mostly by Westerners. By the time of the second and subsequent offerings of this seminar, all teachers were Arabs.[88]

OAPEC is one of the most deeply committed institutions in the Arab world to technology transfer and Arabization.[89] It is by no means finished with what it sees as its task in these areas. Most recently, the OAPEC Secretariat and the joint venture company directors have agreed to set up an Arab engineering venture designed to employ more Arab professionals at the planning and design stages of projects.[90] For the most part, project design and planning have been left to foreign experts, bypassing the growing number of Arab professionals in the region who need work if they are not to emigrate and become part of the brain drain.[91] The success of OAPEC's efforts at technology transfer and Arabization may be seen most clearly in the experience of ASRY. Arab workers have been steadily integrated into every area of ASRY operations since the drydock opened for business.[92] ASRY customers continue to be satisfied, and the drydock operates near or at capacity regardless of the nationality of the workers or the management.

Summary and Conclusions

The agreement establishing OAPEC resembles the charters of the Arab League and OPEC. But along with provisions permitting OAPEC to act as a passive, advisory partner in oil policy decisionmaking dominated by its member governments, the OAPEC Agreement also provides for policy initiation through the active agencies of joint venture companies and a judicial board. OAPEC shares aspects of its passive role with OPEC and the Arab League. Each of these organizations is charged with information exchange and research and development in the energy policy area. In addition, each of the three organizations acts as a forum for energy-policy consensus building. But only OAPEC is empowered by its charter to make decisions in the absence of consensus. This is a reflection of the unusual degree of political and economic harmony that existed among its founding members when the OAPEC Agreement was written and ratified. Part of the original agreement included provisions for a judicial board to settle disputes among the members with regard to any OAPEC activity. When events destroyed the initial harmony among the members of OAPEC, the establishment of the Judicial Board was put off for a number of years.

The lack of political harmony among OAPEC members did not stop the establishment of five joint venture companies. Once created, each company was made independent of OAPEC and the participating member governments. The companies themselves are agents and evidence of regional economic development, and despite the distance carefully established between the companies and their parents, each regards the development of a native industry and work force as a higher priority goal than profitmaking. Training programs were among the first projects undertaken by the companies, which use their own on-site facilities as well as institutions in member countries and abroad for the instruction of personnel. The companies are adequately capitalized, and the member governments have continued to support them in the face of larger than expected operating losses.

The companies are novel international institutions. In spite of the fact that they are owned by member governments, they resemble the joint ventures of private corporations more closely than they resemble national industries. Probably to avoid sovereignty disputes among participating members, the companies are subject to none of the laws of any participating government unless so stated in their charters. In addition, they do not pay taxes in any member state in which they operate. This gives the OAPEC joint venture companies the advantage of operating like national industries across several countries at once.

Technology transfer and eventual technological independence are important OAPEC goals. Along with training programs in the companies and in the new OAPEC Petroleum Training Institute, the organization sponsors seminars and conferences. These, like their counterparts in developed nations, contribute to the professional growth of scientists, technicians, and managers, and create part of the environment in which a native science and technology can flourish.

The joint venture companies and the Judicial Board are institutions that take an active and important role in the integration of the OAPEC member states. They are multinational in the sense that they operate over several separate nation-states. Functionally they are transnational, like the multinational corporations, linking the populations and the economies of the region through educational, commercial, and financial transactions, most of which take place below the governmental level. Together with the more passive integrating functions of OAPEC, they contribute to the community-building capability of the organization.

Notes

1. The Pact of the League of Arab States can be found in an appendix to Robert MacDonald, *The League of Arab States* (Princeton, 1965). A copy of the OPEC Statute (Vienna, 1965) may be obtained upon request from the Public Relations Department of the Organization of Petroleum Exporting Countries in Vienna. The OAPEC Agreement is reprinted in Appendix 1.

2. The relative permeability of organizations is discussed in Nelson Polsby, "The Institutionalization of the U.S. House of Representatives," *American Political Science Review* 62 (March 1968), pp. 145-46. This concept is also used by James G. March and Herbert A. Simon, "The Theory of Organizational Equilibrium," in Amitae Etzioni, ed., *Complex Organizations* (New York, 1965), pp. 67-68.

3. Pact of the League of Arab States, Article 1 (hereafter cited as Pact).

4. OPEC Statute, Article 7-c.

5. OAPEC Agreement, Article 7-b-1.

6. Pact, Article 7.

7. As an example of this kind of outcome, MacDonald described the inability of the league to persuade Saudi Arabia to implement its policies concerning the status and rights of women.

8. OPEC Statute, Article 11-c.

9. Production prorationing refers to the allocation of shares of total oil produced among the members of OPEC. The politics of production prorationing is covered in some detail in Zuhayr Mikdashi, *The Community of Oil Exporting Countries* (Ithaca, 1972), Chap. 5.

10. OPEC Statute, Article 36.

11. OAPEC Agreement, Article 11-c.

12. Ibid.

13. Ibid., Article 11-d.

14. Ibid., Article 11-e.

15. *The Middle East Economic Survey (MEES)*, 7 June 1968, interview with Yamani (unpaged).

16. Ibid.

17. Theodore Lowi, "Four Systems of Policy, Politics and Choice," *Public Administration Review* 32 (July/August 1972), p. 299.

18. OAPEC has jurisdiction over oil matters only. It is not supposed to coordinate economic development beyond the oil sector (Letter from George J. Tomeh, OAPEC advisor, dated 3 March 1978).

19. OAPEC Agreement, Article 25.

20. Ibid., Article 24.

21. OAPEC, *Secretary General's Fifth Annual Report* (Kuwait, 1979), p. 69.

22. OAPEC, *Bulletin*, July 1980, p. 1.

23. Paul Leo Eckbo, *The Future of World Oil* (Cambridge, Mass., 1975), p. 17.

24. The two best examples are Saudi Arabia and Kuwait. See International Monetary Fund, *International Financial Statistics* (April 1978).

25. Tacit collusion refers to an oligopolistic pricing situation in which each seller sets prices according to its perception of the likely pricing policies of the other sellers. See E. H. Chamberlin, *The Theory of Monopolistic Competition* (Cambridge, Mass., 1933), pp. 46-47; F. M. Scherer, *Industrial Market Structure and Economic Performance* (Chicago, 1970), pp. 135-36, 179-82, 443-48.

26. Eckbo, *Future of World Oil*, Chap. 3; Scherer, *Industrial Market Structure and Economic Performance*, Chap. 7.

27. Statement by Yamani quoted in *MEES*, 13 September 1968, p. 7.

28. OAPEC Agreement, Articles 4-6.

29. Ibid., Article 3.

30. Ibid., Article 2.

31. For example, see Amin Saikal, *The Rise and Fall of the Shah* (Princeton, 1980), Chap. 1.

32. *MEES*, 11 October 1958, pp. 1-2, interview with Yamani, quoted on page 2.

33. *MEES*, 24 January 1969, p. 9.

34. Ibid.

35. Statement by Prime Minister Mahmoud al-Magribi to the *Tripoli Mirror*, quoted in *MEES*, 3 October 1969, p. 1.

36. *MEES*, 30 January 1970, p. 7.

37. Reprinted in *MEES*, 13 March 1970, p. 6.

38. For example, the OAPEC Council voted to support Kuwait in its negotiations with British Petroleum and Gulf Oil in 1975-1976 (OAPEC *Bulletin*, December 1975, p. 3).

39. OAPEC also sponsored a seminar on the lube-oil industry in November 1977 as part of its research effort.

40. For example, OAPEC *Bulletin*, February 1978, p. 1.

41. Ibid. Nationalistic investment by OAPEC member countries will be discussed below.

42. OAPEC *Bulletin*, June 1976, p. 3.

43. OAPEC , *Third Annual Report of the Secretary General* (Kuwait, 1976), p. 73; OAPEC *Bulletin*, January 1978, p. 2.

44. OAPEC, *Third Annual Report of the Secretary General*, quotation on p. 74; and interview by author with Dr. George J. Tomeh, OAPEC advisor, 24 April 1978 (hereafter cited as Tomeh interview).

45. For example, a seminar held in Tokyo, Japan on opportunities for cooperation between Japan and the Arab world was sponsored by OAPEC and a joint committee of representatives of several Japanese business organizations. The First Oxford Energy Seminar, held in September 1979, was jointly sponsored by OAPEC, OPEC, and St. Catherine's College, Oxford University, where the seminar was held.

46. Mahmoud S. Amin, "Toward the Creation of an Association of National Petroleum Corporations," OAPEC *Bulletin*, July 1977.

47. *The International Petroleum Cartel*, Staff Report by the Federal Trade Commission (Washington, D.C., 1952).

48. Amin, "Toward the Creation of an Association of National Petroleum Corporations," p. 13.

49. Stephen Duguid, "A Biographical Approach to the Study of Social Change in the Middle East: Abdullah Tariki as a New Man," *International Journal of Middle East Studies* 1 (July 1970).

50. Phillippe C. Schmitter, "A Revised Theory of International Integration," *International Organization* 24 (Autumn 1970).

51. OPEC Statute (Vienna, 1965), Article 2-b.

52. For an exhaustive discussion of the impact of Islam on redistribution, see Maxime Rodinson, *Islam and Capitalism*, trans. Brian Pearce (Austin, Texas, 1978).

53. *MEES*, 25 September 1978, p. 12.

54. Ali A. Attiga, "Regional Cooperation in Downstream Investments: The Case of OAPEC" (paper presented at the OPEC Seminar on the Present and Future Role of the National Oil Companies, Vienna, 10-12 October 1977).

55. *OAPEC Bulletin*, February 1979, p. 5.

56. OAPEC Agreement, Articles 2-e, 5, 28. See Appendix I.

57. One of them, the Arab Petroleum Services Company (APSC), has subsidiaries, such as the Arab Drilling and Workover Company (see Appendix 1).

58. The type of joint venture authorized by Article 5 is different. It could involve outside governments and/or international organizations as well as OAPEC as an entity and/or its individual members as independent participants. Only one OAPEC joint venture has taken this form to date.

59. The company charter specifies the subscribed and authorized capital. The APICORP charter requires a minimum contribution of 3 percent of its capital from every OAPEC member and limits the maximum contribution of each member country to 20 percent unless a greater contribution is "necessary to cover the full capital" (A. Kesmat El-Geddary, "Arab Companies Established by OAPEC," in OAPEC, ed.,

Petroleum and Arab Economic Development [Kuwait, 1978], p. 172, n. 32). Initially, shares in the OAPEC joint venture companies were distributed equally among the members. It is now customary for unsubscribed shares to be distributed equally among the other participants (Geddary, p. 155).

60. The membership of the council acting either as a general assembly or as a supervisory body is limited to oil ministers from participating countries.

61. Geddary, "Arab Companies Established by OAPEC," p. 145.

62. The companies have distinct corporate identities independent of OAPEC and the participating governments. In addition, the charter of each company specifies that it is to operate on a sound commercial basis with the aim of making profits. These qualities make the kind of government participation found in the OAPEC companies a novel approach to national investment.

63. OAPEC Agreement Article 2-e. APICORP is an exception to this rule. See note 59 above.

64. A participating government may act as a custodian by choosing either to keep all its shares or to distribute up to 49 percent of its shares to national corporate entities in which the majority of shares of capital are held by nationals of the participating government relinquishing the shares. However, the government retains its responsibilities to the OAPEC company (Geddary, "Arab Companies Established by OAPEC," pp. 161-62).

65. The AMPTC charter states: "The Company shall enjoy the nationality of all shareholding states" (Article 6); the ASRY charter says, "The Company shall enjoy the nationality of the host country" (Article 6). The host country for ASRY is Bahrain. APICORP and APSC are qualified multinational companies. Article 6 in each of these companies' charters specifies that "The company shall enjoy, with respect to the member states and others, all the rights and privileges of nationality which national companies enjoy in each member state." Because of the novel character of the OAPEC joint venture companies, each has adopted a nationality policy according to its perceptions of what might be most suitable. As the various companies test the validity of their choices of nationality, experience may lead to a unified OAPEC policy in this regard.

66. Geddary, "Arab Companies Established by OAPEC," p. 165. The APSC charter requires only that governments in territories in which the company operates match the exemptions they give to any other company in their country (ibid.).

67. Two chief aims of the OAPEC companies have proven so far to be incompatible. These are the profit-making aim and the mandate of each company to train member nationals for jobs in all phases of the oil industry. Concentration on the second aim has contributed to AMPTC and ASRY operating losses. AMPTC ships are often idle due to lack of business, while the ASRY dry dock has operated at or near capacity since it opened in November 1977. AMPTC continued funding extensive training programs throughout the period of depression in the tanker market and used this as a partial justification for its huge losses—$4.4 million in 1977 (see *MEES*, 5 June 1978, pp. 3-5). ASRY has kept its dry dock operating by price cutting which caused operating losses. The importance of the training program has been offered as a partial justification for the losses of this company also (Interview with ASRY Chairman Sheikh Khalifah bin Salman bin Muhammad in the *Gulf Mirror* [Bahrain] 12 August 1978, reprinted in *MEES*, 28 August 1978, pp. 8-9).

68. Attiga, "Regional Cooperation in Downstream Investments: The Case of OAPEC," pp. 9-10.

69. OAPEC *Bulletin*, October 1977, p. 6.

70. Interview in the *Gulf Mirror* (Bahrain), 12 August 1978 as reported in *MEES*, 28 August 1978, pp. 8-9, quotation on p. 8. ASRY's losses have decreased steadily except during the spring of 1980, when it operated at only 85 percent of capacity in March owing to increased "war risk" premiums assessed on ships entering the Arabian Gulf by their insurers. Capacity was estimated at 92 percent for April and 100 percent for May (*MEES*, 21 April 1980, p. 7).

71. OAPEC *Bulletin*, March 1978, p. 5. During its first year of operations (9/23/77-9/23/78), ASRY repaired 100 ships according to a report in the OAPEC *Bulletin* of November 1978. The vessels repaired are listed by name and carrying capacity only, so it is impossible to say definitely what percentage of them were Arab vessels. From an inspection of the published list, I would estimate that fewer than 10 percent of the vessels repaired were Arab ships, a testimony to the attractiveness of ASRY's services and rates.

On 27 March 1980 the Arab Center for Coordination and Maritime Consultations, on behalf of AMPTC and several national tanker companies, signed a two-year contract for repair and maintenance with ASRY. The contract was made to reduce the costs of maintenance and repair to the tanker companies (*MEES*, 7 April 1980, p. 8).

72. Interview with ASRY chairman published in OAPEC, *Bulletin*, December 1978, pp. 1-3.

73. Ibid., p. 3.

74. Attiga, "Regional Cooperation in Downstream Investments: The Case of OAPEC," p. 15.

75. Nur El-Din Farraj, "Highlights of APICORP's Activities During 1978," Special Report to OAPEC *Bulletin*, February 1979, p. 21.

76. Ibid., pp. 18-22. Among APICORP's loans for that year were a second loan to the Qatar Petrochemicals Company of $175 million for an ethylene project, and $117 million to Sonatrach (Algeria) to finance a new LPG pipeline (OAPEC *Bulletin*, December 1978, p. 3).

77. ADWC is owned by APSC (40 percent), by APICORP (20 percent), and by Santa Fe International Services, Inc., an American firm (40 percent). See *MEES*, 10 March 1980, p. 11. Within eighteen months of its organization, ADWC was operating seven drilling rigs in Libya.

78. Farraj, "Highlights of APICORP's Activities During 1978."

79. *MEES*, 22 May 1978, p. 4.

80. Ibid.

81. OAPEC has eight crude oil carriers and two LPG ammonia carriers. AMPTC purchased two LPG tankers in 1979 (OAPEC *Bulletin*, June 1979, p. 4) and at its thirty-third board meeting in May 1980 began plans to expand the fleet still further (ibid., June 1980, p. 1). These represent additions to the fleet of eight crude carriers.

82. OPEC, *Annual Report-1978* (Vienna, 1978), Table 1.

83. Ibid., p. 90.

84. These amounted to SR 23.9 million in 1976; SR 41.8 million in 1977; SR 61.5 million in 1978; and SR 100 million in 1979. OAPEC *Bulletin*, December 1979, p. 6, (June 1979), p. 4. (April 1980), p. 5. Amounts are in Saudi Riyals.

85. For a discussion of these and other dependency problems resulting from direct foreign investment by multinational corporations, see Theodore H. Moran, "Multinational Corporations and Dependency: A Dialogue for Dependentistas and Non-Dependentistas," *International Organization* 32 (Winter 1978).

86. OAPEC *Bulletin*, April 1980, p. 2.

87. Such as acidizing and cementing and hydrocarbon logging. OAPEC *Bulletin*, March 1979, p. 3.

88. Tomeh interview.

89. Secretary General Ali Attiga presented a paper on the urgent need for technology transfer at the Conference on Opportunities for Cooperation Between France and the Arab World held in Paris in November 1975. Editorially, the *Bulletin* has pressed for greater concentration in the area of technology transfer (for example, see the October 1976 issue).

90. This company, the Arab Engineering Consulting Company, will have the national oil companies of the member governments themselves as shareholders. See Chapter 6 below and *MEES*, 23 June 1980, p. 7.

91. OAPEC *Bulletin*, December 1979, p. 1. The need for such an effort is described by A. B. Zahlan, *Science and Science Policy in the Arab World* (New York, 1980), Chaps. 1-2.

92. "... [C]ompany sources report that ... Arab nationals constituted 52% of ASRY's work force of 1321 and 50% of its senior staff. A total of 686 employees underwent training locally, while 28 were sent abroad. ... " OAPEC *Bulletin*, February 1980, p. 1. The AMPTC Board agreed to a ten-year plan for Arabizing the tanker company's workforce in February 1976 (OAPEC *Bulletin*, April 1976, p. 2). However, information about AMPTC trainees has become increasingly scarce since then, both in the *Annual Report of the Secretary General* and in the monthy OAPEC *Bulletin*. AMPTC was not even listed in the bulletin's index for 1978. All the other joint companies were listed. It is as though the entire organization is uncomfortable with AMPTC's financial situation (see next chapter) and prefers not to examine anything about the tanker company too closely.

Chapter 4

OAPEC and Arab Community

Investigations into the nature and extent of Arab community are increasingly faced with the problem of defining and explaining the changing role of Pan-Arabism.[1] As discussed in Chapter 1, Pan-Arabism is made up of four factors: shared religion, language, and culture across the Arab world; the history of colonial exploitation, including a sense of having been betrayed by the nations of the developed West, particularly Great Britain; the problems most of the new national governments had in establishing their political legitimacy; and, finally, the collective Arab decision with regard to the illegitimacy of the state of Israel and their generally consistent refusal to accommodate its existence in the region. In addition to the series of wars between Arabs and Israelis since 1948, the Arab Economic Boycott is a continuing manifestation of this feature of Pan-Arabism.

The most drastic change in Arab regional politics in the 1970s has been the development by Egypt of an independent foreign policy with regard to Israel. Egypt's decision to sign a peace treaty with Israel set off reactions in individual Arab states and in Arab and Islamic regional political and economic organizations. Arab reaction was so strong that Egypt was suspended from membership in these organizations, including OAPEC and the three OAPEC joint venture companies in which Egypt had shares. Arab institutional community, to a very large extent, failed to override Arab communal political impulses. The small saving grace in the formal Arab response to the 1979 peace treaty was its careful description of suspension as a temporary condition, one that did not affect Egypt's status as "part of the Arab nation and the Arab homeland, both as a country and as a people."[2] And although the institutional ostracism of Egypt proceeded rapidly during the spring and summer of 1979, most structural ties, with the important exception of trade, were maintained. This outcome is interesting in the context of neofunctionalist theories of regional integration as it demonstrates the independence of structural and institutional relationships.

Even without considering the events in Egypt, Arab community is in an unsettled state. In the past ten years, for example, there have been more than a dozen instances of armed conflict between Arab countries, including conflict between OAPEC members.[3] Arab countries have often united and divided according to ideological commitments to revolution or traditionalism. The formation of the United Arab Republic and the on-again, off-again relationship between Iraq and Syria are cases of this. Yet there have been instances of revolutionary-monarchist alliances as well, the most recent being the rapprochement between Iraq and Saudi Arabia after the Iranian revolution and the Egyptian-Israeli peace treaty.

The assumption in this chapter is that there are more factors uniting the Arab countries, particularly the Arab oil exporters, than dividing them. Evidence of modern Arab community will be sought using the eclectic community model developed in Chapter 1. The results of a series of empirical analyses and a brief look at Arab interdependence through labor migration will be used to examine the nature of structural community among Arab countries. After this, institutional community within OAPEC will be discussed and evaluated according to standards developed by students of regional integration. Finally, some of the questions posed in Chapter 1 about the nature of Arab community and OAPEC's role in it will be addressed. Could OAPEC create, by itself, a sufficient basis for Arab community? Does OAPEC inhibit or encourage cooperative international relations in the Arab world? Does OAPEC complement or compete with other Arab international organizations, principally, the Arab League? Has OAPEC materially aided the progress of economic development in the region? How might OAPEC serve as a model for other regionally-based commodity organizations interested in economic development and the preservation of member state autonomy in the management of their resources?

Structural Arab Community

The structural community model describes transnational interstate relations, that is, relationships below the governmental level between individuals and groups in different countries. These kinds of relationships, such as trade, exchanges of factors of production, communications, and transnational identification, are thought by some, most notably the functionalist and neofunctionalist schools of regional integration theory, to inhibit international conflict and to promote international cooperation. Structural community has been developed as a two-tier model. One level includes international transactions and the other, the interdependence resulting from these transactions and from other integrating factors operating on a given group of states. "Transaction" is broadly defined. The term includes perceptions of attribute similarity (exchanges of recognition) as well as behavior such as trade or communications (exchanges of goods or messages). This level of structural community can be compared to the transnational society concept of Karl Kaiser[4] or to the horizontal interdependence concept developed by Richard Rosecrance and his associates.[5] Kaiser's transnational society is limited to elite-level transactions. Rosecrance's horizontal interdependence refers to transactions at the mass level as well and is more similar to the structural community model used here.

The second level of the structural community model might be called the collective effect of the first. Transactions in the broad sense defined here create pathways that transmit unintended effects as well as intended ones. Exchanges of goods, capital, labor, communications, and recognition can also transmit inflation, unemployment, and international terrorism. This might show itself as a response by a sensitive partner to events occurring in another country to which it is linked via a transaction network (and perhaps also by an organizational regime, which will be discussed later). This level of structural community can be compared with the Rosecrance concept of vertical interdependence[6] or to the sensitivity component of international interdependence as defined by Robert Keohane and Joseph Nye.[7]

Arab structural community will be discussed here using the results of empirical techniques designed to uncover transaction patterns. Interdependence itself is very difficult to operationalize and at this point virtually impossible to measure quantitatively in developing countries using empirical methods.[8] However, inter-Arab migration will be examined as an example of such interdependence.

Attribute Similarity

One way to infer that structural community exists between peoples is to examine common attributes that might cause them to perceive that they share important national characteristics with their counterparts in other countries. According to Albert Hourani, the modal "Arab" is an ethnic Arab Sunni Muslim.[9] If we divide the Arab countries into two groups, one containing the ten OAPEC nations and the other containing ten members of the Arab League that are not members of OAPEC, the groups can be compared regarding their ethnic and religious composition (see Appendix 2). When this is done, we find that both groups have about the same percentage of Sunni Muslims but that the group of OAPEC countries has a significantly larger percentage of Arabic speakers and Muslims of all sects. Hourani felt that ethnicity was only marginally important to the feeling of "Arabness" in the minds of people living in Arab countries, but that monosectarian adherence to Islam was the critical factor supporting political cohesion within Arab countries.[10] This judgment is supported by events in Lebanon which led to civil war in the 1970s, where the internal factions split primarily along religious lines. Sectarian rivalry can also be seen in non-Arab Iran, where the Shi'ite revolutionary forces have fought ethnic Sunni Arabs demanding regional autonomy in southwestern Iran.

Since the Iranian revolution generally, Westerners have become more sensitive to the power of Islam and ethnic identification to unite or divide peoples throughout Africa, Asia, and the Middle East. Iran itself has been a microcosm of such behavior. Shi'ia clergy led the successful Iranian revolution, which has since been beset by ethnic and sectarian separatist movements variously composed of Kurds, Baluchis, Azerbaijanis, and others. Ethnic and religious identification is very important in the region and evidence of higher levels of religious and ethnic homogeneity in the set of OAPEC countries than in the set of non-oil-exporting Arab states indicates a somewhat higher potential for international relationships relatively free of religious and ethnic conflicts.

Another way to look at similarities among several nations is to group them according to how alike they are when sets of attributes are considered. According to this method, OAPEC countries tend to be more like each other than like most other Arab countries when religious and ethnic factors are considered simultaneously. If the same procedure is applied to indicators of economic development, this neat separation does not occur. OAPEC countries are among the richest and the poorest of the countries in the region and their individual levels of industrialization are similarly diverse. Similarly, when tariff structures and trading patterns are compared, there is a split in OAPEC members. This split, however, is regional. OAPEC countries in North Africa have very different trading patterns, but the five OAPEC nations of the Fertile Crescent and the Arabian Peninsula are similar except for varying degrees of dependency regarding the purchase of arms.

Finally, when levels of militarization are used to group the Arab countries, we find a separation between three big spenders, Egypt, Saudi Arabia, and Iraq, and the rest of the Arab League including the remaining OAPEC members. Even the big three differ from one another rather than forming a single group. Only oil-exporting Arab countries have money enough to afford large military expenditures, but these countries buy their arms from different sources and show varying levels of dependence on the United States and the Soviet Union. Even without additional funds from Saudi Arabia, Egypt has continued to support an army large enough to stand as a threat to any other nation in the region.[11]

Although military expenditures in the region are increasing over all owing in part to increases in national incomes enabling countries to spend more money on arms,[12] only the big three countries have chosen to militarize to the extent that they can be regarded as actual or potential military powers on the level of Iran or Israel. Even though few Arab oil-exporting countries can be regarded now as regional military powers, their oil incomes allow one to think of some of them as potential military powers or as financiers of regional military powers.[13]

So far, the best that can be said about Arab community in OAPEC as opposed to the Arab League as a whole shows only potential bases for solidarity among OAPEC nations or among Arab countries generally. In part, this is owing to overlapping memberships in the various groups showing similarities. A given OAPEC country is not always similar to the same other nations when different characteristics are considered. In addition, nations that do group together are not exclusively oil-exporting or oil-importing Arab countries. Whatever similarities bind oil exporters together bind them as closely to Arab League members without significant oil resources. Although one cannot infer that OAPEC is a better base for Arab community than the Arab League as a whole, it is also true that it does not appear to be inherently divisive. Its apparently neutral role may be one result of institutional changes within OAPEC that brought it into closer conformity with the expectations of the rest of the Arab League.

Patterns of Trade

Trade is the only transaction variable for which reasonably complete data is available, and even in this case, some of the data is very old. Some question exists regarding the reliability of the data, and there is a problem in comparing trade statistics from Arab OPEC countries, which all show large amounts of exports going to developed countries, with those of other nations in the region. In spite of this, we can see that a few Arab nations do trade extensively with their neighbors; in addition, commodity dependence (a high percentage of a country's imports or exports being of one commodity) does not appear to lead to partner dependence (a high percentage of a country's trade being with one partner) in either imports or exports for most Arab oil-exporting states. Import dependence for capital and consumption goods is usually generalized to several developed countries rather than being specific to one. This allows Arab importers to minimize their client roles in relationships with individual developed countries.

Table 6 shows the pattern of trade for twenty countries of the Arab League in the most recent year for which data from the same source was available. On the whole, countries with significant concentrations of trade going to their regional neighbors are the non-oil Arab states that import oil, usually in exchange for agricultural products. Oil products are also occasionally trans-shipped between oil-exporting countries, but usually not in proportionately large amounts. For example, Bahrain is an oil exporter that now exports more to regional partners than to outsiders. Most of its exports are manufactured products, particularly petrochemicals. Bahrain obtains the majority of its imports of capital and consumption goods from Western Europe, the United States, and Japan, and cannot be said to be specifically import dependent, although it is generally dependent for its imports on the Organization for Economic Cooperation and Development (OECD) countries.

Iraq is an example of a country that has diversified its exports so that no single recipient receives more than six percent of the total value of its exports.[14] Nearly half of Iraq's exports are mineral fuels, but most of the value received for these exports is for processed materials, increasing Iraq's oil

TABLE 6 Percentage of the Value of Arab Country Imports and Exports by Various Countries and Regions for the Most Recent Year

	Algeria[1,A,B,C]	Bahrain[1,B]	Egypt[1,A,B,D]	Iraq[2,A,B]	Jordan[1,A,E]	Kuwait[2,F]	Lebanon[3,A,G,H]	Libya[2,A]	Mauritania[3,I]	Morocco[1,A]
Middle East										
Imp.	0.2	46.8	3.2	2.1	17.9	1.9	10.1	1.2	0.3	7.0
Exp.	n.a.	30.4	8.1	n.a.	71.7	6.6	50.4	2.8	0.9	2.2
North Africa										
Imp.	0.6	0.0	0.7	0.6	2.5	0.2	1.6	0.8	1.4	0.1
Exp.	0.0	0.1	3.6	n.a.	2.5	0.5	12.6	0.1	0.0	1.6
Western Europe										
Imp.	67.6	22.0	47.9	55.1	40.8	37.8	54.6	71.9	67.1	65.1
Exp.	43.3	0.8	37.9	n.a.	6.3	31.8	14.2	60.0	63.1	69.2
Eastern Europe										
Imp.	4.8	0.1	12.3	8.5	7.2	0.7	8.8	6.0	0.2	4.9
Exp.	1.8	n.a.	39.6	n.a.	4.9	n.a.	5.6	2.6	n.a.	9.5
United States										
Imp.	8.8	6.6	16.4	5.3	14.8	14.9	12.0	4.1	9.5	6.3
Exp.	48.4	8.6	2.0*	n.a.	1.9*	0.6*	5.4	27.2	2.8*	2.2*
Developed Asia										
Imp.	6.2	8.6	5.2	13.9	6.3	21.3	4.0	8.3	0.5	2.9
Exp.	0.4	15.7	3.6	n.a.	3.2	22.2	0.3	2.6	9.7	1.8
First Partner										
Imp.	23.1[a]	45.1[c]	16.4[b]	21.2[c]	14.8[b]	21.3[f]	12.0[b]	25.5[g]	50.0[a]	27.3[a]
Exp.	48.4[b]	15.8[c]	23.2[d]	n.a.	29.8[c]	22.2[f]	19.4[c]	27.2[b]	15.3[h]	24.7[m]

Table 6 continued

	Oman[2.B]	Qatar[1]	Saudi Arabia[2]	Somalia[4]	Sudan[2]	Syria[2.A.J]	Tunisia[2]	UAE[2]	ARY[2.D.K]	PDRY[5.A]
Middle East										
Imp.	22.8	10.7	24.5	6.4	11.7	16.2	7.9	8.4	16.0	39.9
Exp.	0.2	n.a.	2.8	79.3	3.1	9.4	3.1	2.0	40.3	6.8
North Africa										
Imp.	n.a.	0.1	0.6	2.2	1.1	1.7	1.0	0.1	0.7	2.5
Exp.	n.a.	n.a.	0.4	0.2	3.6	2.7	4.4	0.0	n.a.	3.5
Western Europe										
Imp.	37.5	44.5	32.0	51.3	51.1	50.7	71.3	38.9	25.9	15.8
Exp.	23.5	n.a.	41.0	8.3	53.3	65.2	51.8	45.1	21.9	31.6
Eastern Europe										
Imp.	0.6	0.4	0.9	8.4	0.6	10.1	3.2	1.9	3.3	2.2
Exp.	n.a.	n.a.	0.1	5.5	1.3	14.0	3.7	1.2	n.a.	n.a.
United States										
Imp.	6.1	9.6	19.3	2.4	9.4	6.9	6.2	13.4	3.8*	1.3*
Exp.	15.8	n.a.	4.8	0.1*	3.9	1.0*	13.8	11.8	0.5*	0.2*
Developed Asia										
Imp.	12.1	26.7	12.6	0.6	6.5	7.3	1.5	17.4	10.2	13.1
Exp.	43.2	n.a.	20.1	0.0	7.5	0.4	0.2	27.6	0.6	10.5
First Partner										
Imp.	19.1[h]	26.7[f]	19.3[b]	32.0[g]	20.3[h]	13.9[c]	32.2[a]	17.4[f]	11.6[c]	14.9[j]
Exp.	43.2[f]	n.a.	20.1[f]	64.1[c]	19.7[g]	15.3[g]	21.5[g]	27.6[f]	33.4[i]	21.9[h]

Source: United Nations, *Yearbook of International Trade Statistics 1977*, Vol. 1 (United Nations, 1978).

1 1977
2 1976
3 1973
4 1975
5 1969

* Includes Canada.

a France
b United States
c Saudi Arabia
d USSR
e Federal Republic of Germany
f Japan
g Italy
h United Kingdom
i China
j Iran

A Excludes some or all military equipment.
B Excludes some or all precious metals.
C Excludes crude oil exports.
D Excludes foreign aid.
E Excludes trade of concessionaires.
F Excludes data from the Neutral Zone.
G Excludes data from Palestinian relief efforts.
H Excludes oil imports, except for what is consumed internally.
I Excludes smuggled goods.
J Excludes trade of international organizations.
K Excludes imports for government projects.

income by the value added for processing, and decreasing its export dependence even further.

Some interesting patterns are revealed by Table 6. The African Arab countries, with the possible exceptions of Egypt and Somalia, are more trade dependent than other Arab countries. Most of these countries are heavily dependent on one or more European nations, chiefly France and Italy. However, extreme export dependence is not associated with developed country trade nor with oil exporting per se. Although rates for import dependence are higher than for export dependence, these too are independent of oil exporting, although in some cases they are not independent of trade with a single developed country.

Perhaps the most surprising result of an examination of Arab trading patterns is the diversity of import sources and export destinations for the region as a whole and for many individual countries as well. Most Arab countries have trade ties to the United States and, to a lesser extent, to Eastern Europe. Extensive ties to Western Europe and Japan provide alternative sources of technology and consumer goods independent of the deteriorating relationship between the two core powers. The former French and Italian colonies still maintain very close trade ties to the former imperial powers. It seems to be colonial heritage rather than commodity dependence that exerts this disproportionate influence on Arab trade patterns.

One can approach the study of inter-Arab trading patterns in a different way. Using time series data on the intra-regional and extra-regional trade of some Middle Eastern countries, Barry W. Poulson and Myles Wallace showed, in a 1979 article in *The Middle East Journal*, that the bulk of intra-regional trade was among the Asian countries, that is, among the Fertile Crescent and Arabian Peninsula countries. Their results are similar to those in Table 6 for one year's trade.

Poulson and Wallace noted a big jump in intra-regional trade in 1973 and 1974, which has since fallen in proportion to extra-regional trade for these same countries. They conclude that regional integration through trade has increased, even though there is some small evidence of a relative decline. However, they regard the evidence of extensive trade with Europe, Japan, and North America as signifying an increase in economic dependence. This conclusion is significantly different from the one reported above.

Dependence can be defined in a number of ways. Poulson and Wallace imply that dependence is reliance on outsiders for significant amounts of imported goods.[15] I have called this "generalized trade dependence." However, countries that are normally thought of as powerful, as *interdependent* at best (meaning that their relations with other countries operate on a basis approaching equality), are equally trade dependent, usually through commodity dependence rather than generalized trade dependence. For Japan and for France, if not so much for the United States, commodity dependence has led to modifications in foreign policy toward the Middle East, most notably in these nations' postures toward Israel. In a sense, the generalized trade dependence of oil-exporting countries gives them more freedom to maneuver than the commodity dependence of their trading partners. Most regional development projects are models of internationalism, with multinational companies bidding on contracts and projects taking shape through the efforts of several of them rather than only one.[16]

There is great pressure on the multinational development corporations from their governments to win contracts in oil-exporting countries. This is part of the "recycling" process, the funneling of oil country financial surpluses back to the oil-importing developed countries.[17] The most extensive picture of the competitive behavior of multinationals trading in the Middle East has emerged from investigations into bribery relating to arms sales,[18] which demonstrates that the dependence characterizing trade relationships between developed and oil-exporting developing countries is mutual and not one-sided. If Arab dependence on the outside has increased, so has extra-regional dependence on Middle Eastern oil exporters. This mutual or codependence pushes such international relationships back toward relative equality.

Whatever the nature of extra-regional trade, it does not detract from the significance of increases in trade within the Arab world. Oil exports distort the importance of trade among Arab countries. However, the current boom in development projects in the region as a whole, funded by oil exporters directly or through foreign aid (see below), will not last. In spite of the unfortunate behavior of many of the oil rich countries, which spend for prestige as much as for economic development, there are only so many projects that one country can assimilate. Once this level is reached, or once the governments perceive that a lot of their money is being wasted, imports of material and labor from the developed countries will decline. The trade with developed countries, which appears most important now for Arab oil-exporting countries, will become the least important for many of them in the long term. As the governments of the Arab countries fear, when the oil and the money are gone, intra-regional bonds are likely to be the only ones left.

Inter-Arab Migration

There are two major reasons for the extensive inter-Arab migration which has occurred over the last generation. One reason is the displacement of millions of Palestinians after the establishment of the state of Israel. Palestinians are among the most highly educated Arabs. They have dispersed throughout the region, many holding important jobs in their adopted countries. According to data obtained by several Palestinians in an ambitious study undertaken in 1969-70 to locate exiled Palestinian university graduates,[19] Palestinians were found to be distributed as shown in Table 7. Since this research was completed, the influx of greater and greater amounts of oil money has increased the concentration of Palestinians and other migrants in oil rich Arab countries.

TABLE 7 *Location of the Palestinians, 1969-1970*

Jordan	900,000
West Bank	670,000
Gaza	364,000
Israel	340,000
Lebanon	240,000
Syria	155,000
Kuwait	140,000
Egypt	33,000
The Gulf*	15,000
Libya	5,000
Saudi Arabia	20,000
United States	7,000
Latin America	5,000
West Germany	15,000
Total	2,923,000

Source: Nabeel Shaath, "High Level Palestinian Manpower," *Journal of Palestine Studies* 1 (Winter 1972), reprinted by the Arab Information Center (New York, n.d.).

*Includes the Emirates, Bahrain, and Qatar.

Palestinians continued to value education after the diaspora began in 1948. Perhaps in response to their values, Palestinian university graduates have concentrated in educational professions. Large numbers of Palestinians populate the faculties of schools in Kuwait, Saudia Arabia, Algeria, and Libya.[20] The next largest concentration of educated Palestinians is in the engineering profession, and the third largest in medicine. Palestinian engineers and doctors are especially important to the Arabian Peninsula countries, where they are involved in various economic and social development projects.[21]

While Palestinians are important additions to the educated populations of many Arab states, they are also perceived as political risks, as timebombs waiting to explode and destroy the remaining traditional monarchies that control the richest Arab states. Palestinians are intimately involved in radical Arab nationalist and Palestinian organizations. They have become a strong force in international politics generally since the status of the Palestine Liberation Organization (PLO) and its leader, Yassir Arafat, rose in the 1970s. Palestinians mounted a revolution of sorts in Jordan in 1970, which was suppressed in very bloody fighting. They are also combatants in the Lebanese Civil War. Fluctuations in the relative power of Palestinian militants in Lebanon have led Syria to alternate its support from one side to the other in its attempt to maintain the Palestinians as threats to Israel and not to itself. Ambivalent feelings toward Palestinians by other Arabs support the maintenance of refugee camps in Lebanon and elsewhere in the region alongside a willingness to employ Palestinians in important positions in many Arab countries. This has caused the ambiguous position of the Palestinians to remain a problem for more than one nation in the region.

The other impetus toward inter-Arab migration is the growth in oil revenues, which has enabled oil-exporting Arab countries to increase the pace of their economic development efforts and thus to create demand for skilled and professional workers in countries with small populations and large incomes. Egypt and Jordan are the two nations that have contributed the largest numbers of skilled workers to the migration. The situation of Egypt will be examined here because it demonstrates very clearly the benefits and the costs of international interdependence in Middle Eastern labor markets.

There are several reasons why Egyptian workers emigrate. There is widespread unemployment and underemployment in Egypt. Over 11 percent of the work force is currently unemployed. The nation's high rate of population increase means that, if unemployment is not to increase further, over a third of a million new jobs must be created every year.[22] A related reason for emigration is the difference in wage rates in Egypt from those in richer Arab oil-exporting countries. Changes in the rate of labor migration appear to be a function of changes in relative wage levels in the labor-exporting and labor-importing Arab countries. There is even wage competition among labor-importing Arab countries, due to the increased demand for labor that occurs when oil revenues increase.[23] High wages as well as the opportunity to work at jobs that might be foreclosed to individuals owing to the rigidity of Egypt's domestic labor market are the main incentives to emigration.[24]

The pool of labor from which Egyptian emigrants come consists of urban workers who do not have government jobs. There is little or no migration of agricultural labor out of the country or within Egypt from rural to urban areas.[25] Urban workers with government jobs have more reasons to stay in the country than to leave. Their jobs provide secure, if low, incomes; they are usually able to get second jobs to increase their incomes; and annual raises are nearly automatic for government workers, regardless of performance.[26]

Potential and actual emigrants from Egypt are estimated at about 1,400,000 workers.[27] This is a small number compared with the size of the work force itself, estimated at nearly 13,000,000 persons in 1976, including 600,000 persons already working abroad and 1,479,000 persons unemployed inside Egypt.[28] There are several reasons for the disparity. First, skilled or experienced workers are in greatest demand in Arab labor-importing countries. Second, what we would call "hard core" or structurally unemployed Egyptians, those who are inexperienced, unskilled, and unemployed, cannot afford the cost of passports and travel, and the kind of work they would be eligible for if they were to migrate is relatively poorly paid because of competition from unskilled Asian labor.[29] What Egypt is doing, then, is exporting its small pool of skilled non-government workers.

The Egyptian government encourages labor migration. Labor export reduces domestic unemployment and brings in foreign exchange to assist in financing purchases from abroad. Remittances now amount to approximately $2 billion per year.[30] Egyptian workers abroad also send consumer goods back into Egypt.[31] This, however, is regarded as a mixed blessing.

The distribution of Egyptian workers abroad is shown in Table 8. For the most part, Egyptians work in countries close to their own. The labor shortage in Oman has not attracted very many Egyptians because of the greater traveling distance.[32] By far the greatest number of Egyptians work in Libya and in the rich oil-exporting countries of the Arabian Gulf. Interestingly enough, however, 4,500 Egyptian workers have emigrated to Jordan, another net labor exporter in the region. This is called "replacement migration." It is highly significant because it means that Jordan is able to outbid Egypt for labor to alleviate domestic shortages of workers caused by Jordanians leaving to work abroad.

TABLE 8 *Location of Egyptians Abroad, 1975-1977*

Country of Residence	Year	Number in Thousands	Percentage of Egyptians Abroad
Libya	1976	275.0-380.0	59.6
Saudia Arabia	1975	110.0-130.0	20.4
Kuwait	1975	60.5	9.5
UAE	1975/6	22.0	3.5
Qatar	1976	16.2	2.5
Sudan	1973	11.8	1.8
Jordan	1977	4.5	0.9
Bahrain	1976	3.5	0.5
Oman	1975	1.9	0.3
Other Arab States	1975	4.0-5.5	0.9
Total		510.9-637.4	100.0

Source: J. S. Birks and C. A. Sinclair, "Egypt: A Frustrated Labor Exporter?" *The Middle East Journal* 33 (Summer 1979), Table 5, p. 295.

When other Arab countries began imposing sanctions on Egypt for having signed the 1979 peace treaty, the sanctions did not include refusals to accept Egyptian migrant labor. The role of Egypt as supplier of labor to the region is very important, much more so than what has been judged by the other Arab governments as a temporary situation.[33] This is an example of the importance of structural linkages in international relations. In this, as in other cases, structural ties are so important to both parties in certain international transactions that violent political disagreements do not disrupt them.

Although the importation of Egyptian workers has been very good for the development of industry and infrastructure in rich Arab countries that have used imported labor to carry out ambitious development projects, the effects of labor migration on Egypt have been mixed at best. Egyptians who migrate tend to be the most skilled workers, teachers, and technicians. This reduces the pool of skilled workers left in the country to carry out Egypt's own development projects. There is a particularly acute shortage of teachers in Egyptian universities, which must sustain high student-teacher ratios, while the new universities in Arab oil-exporting countries have large faculties and few students.[34] Another problem is that because high wages abroad make the rewards for migrating so much greater, Egyptian workers at home are thought to be discouraged from increasing productivity and working for advancement.[35]

A growing social problem associated with Egyptian labor emigration is the transfer of consumption patterns from rich oil states to the poor domestic economy. Although workers abroad do return a large portion of their wages as cash remittances, they also send home consumer goods, which stimulate an already present Egyptian tendency to consume beyond the nation's means. Ibrahim Sa'ad Eddin writes of Egyptian urbanites going through great hardship in order to be able to afford luxury goods, including renting their homes in fashionable areas to tourists while they live in less expensive sections so they can spend the difference.[36] The money and luxuries that stimulate consumption reinforce the stimulus to consume which rich Arab tourists provide through their examples. Tourism, too, has continued since the peace treaty. Egypt is centrally located, has an Arabic-speaking population, and a more relaxed lifestyle than the rigors of the Shari'a allow in countries such as Saudi Arabia. The rush of oil money to OPEC nations has stimulated both their own consumption and that of the poorer countries with which they are interdependent.

Although researchers in Arab countries and in the West are beginning to write about the social and economic costs to Egypt of skilled labor emigration, the government is continuing to encourage more workers to go abroad.[37] There is some hope that replacement migration will begin eventually to tap workers in that hard-core category of the Egyptian unemployed. Remittances from abroad have increased so substantially since 1973 that the government is willing to accept some of the domestic costs of labor migration in order to achieve the foreign exchange benefit. The export of Egyptian labor is also a potential source of power to the Egyptian government in its relations with other Arab countries, should such relations resume. Labor-importing countries need the workers they get from Egypt. Their development projects would be farther from completion without the productivity of Egyptian labor. In order to continue to tap that labor source, the Egyptian government might require the other Arab governments to stop discriminating against Egyptians in wage rates, to train some unskilled Egyptian workers along with employing skilled Egyptians, and to deduct and transfer a portion of the wages of each Egyptian worker to the home government.[38] The last provision would help the government to curb domestic consumption of luxury goods and would force Egyptians abroad to take some of the tax burden from those who cannot afford to migrate.

The exportation of labor by Egypt and other low-income Arab countries such as Jordan and Yemen is an example of international interdependence in the Middle East. Structural ties from labor-exporting to labor-importing countries are difficult to break. Labor-exporting countries need the foreign exchange their workers earn abroad, and labor-importing countries cannot implement their development plans without the help of foreign workers. Unfortunately, in the case of Egypt and perhaps of other labor-exporting countries, the economic benefits of labor migration are offset by domestic social and economic costs. Transnational links created by its workers abroad transmit more than income back to Egypt. They also supply role models for domestic consumption and luxury goods from abroad. The high wages that lure Egyptian workers to Libya and the Arabian Gulf states depress productivity in Egypt and drain the country of needed skilled labor. In spite of this, Egypt continues to send workers abroad. While the country is isolated from other Arab nations in the international organizations from which Egypt has been suspended, its migrating workers constitute an important connection between Egypt and the other Arab States.

Palestinian workers are another source of structural linkage in the Arab world. The Palestinians, being stateless, are scattered throughout the region and beyond. Palestinian ties to one another create a transnational network that has supported terrorism in the region as well as the movement of skilled workers to labor-short Arab states. The migration of labor in the Middle East links Arab countries in important ways. Breaking ties formed by the movement of labor between Arab countries would create extensive disruption in the economies of rich and poor Arab states alike. Although institutional community ties may be broken in response to political behavior by governments, structural ties are less easily ruptured because of the high domestic costs all parties incur when this happens.

Summary

Structural linkages in the Arab world are not confined to those among Arab oil-exporting countries. When Arab countries are grouped according to their similarity on one attribute or a set of related attributes, the oil-exporting countries neither form a group of their own nor form groups in which only oil exporters are members. Even when variables are chosen that indicate levels of economic development, Arab oil-exporting countries are not located at similar points on continua ranging from primitive to modern economic characteristics. Indeed, before the outbreak of civil war in 1975, Lebanon might have been called the most economically developed Arab country, and Lebanon has no oil. Libya, a nation that has a great deal of oil and gets more money for it than other OPEC countries, is among the least economically developed Arab countries. Thus being an oil exporter does not remove any given Arab country from similarity to and thus community with other Arab countries.

By the same reasoning, Arab oil exporters are not cut off in politically and economically relevant ways from the rest of the Arab League. They share economic, historical, religious, and other characteristics with countries that do not have their oil wealth. It has been a goal of Arab oil-exporting countries *not* to appear as a bloc within the Arab League. The analysis here shows that they have retained many similarities with their neighbors, and that their national interests as a whole are unlikely to diverge from those of non-oil Arab countries. Characteristics such as regime type, which continue to divide Arab nations, cut across their division into oil-exporting and oil-importing groups.

A major source of structural community among Arab nations is the extensive migration of Arab workers from one country to another. This migration is possible because of the attribute similarities, particularly religion and language, that we have examined before. The migrants discussed in this chapter, Palestinians and Egyptians, are not the only Arab migrants. Migrants come from Yemen and Jordan in numbers nearly at large. The Palestinians are unique in their influence on Arab countries because they are part of a national diaspora that began with the establishment of the state of Israel in the Palestinian homeland. Palestinians are an important addition to the labor forces of other Arab countries because they are highly educated. They are also a constant reminder of one of the most powerful links among Arab countries, their vow to isolate if not to eliminate the state of Israel from their region. Even though Egypt has broken ranks with the rest of the Arab League with regard to Israel, Arab resistance continues.

Skilled Egyptians also participate in inter-Arab migration. The case of Egypt shows how migration hurts a home country which is itself short of skilled workers. The interdependence of Arab states because of national attributes and national policy create, to some degree, a single labor market for some kinds of workers in the Arab world. This joint labor market enables rich Arab countries to outbid others for the labor of nationals of low-income countries. Labor migration and the high salaries that support migration in turn have an impact on consumption patterns and tax revenues in the home countries. A common labor market, like a common trading market, exploits the economically worse off members of the community by diverting economic activity that might otherwise have been stopped by national boundaries from poor countries to rich ones.[39]

Egypt is in a particularly poor position to be a part of a common Arab labor market. When its relations with other Arab countries were broken off in 1979, Egypt lost its ability to bargain with labor-importing Arab countries for equal treatment of its nationals working there. It also lost the ability to bargain for benefits other than employment for its nationals and remittances to the home country. As an important supplier of labor to other Arab countries with better training facilities, Egypt could have asked for training for some of its unskilled workers in exchange for keeping its borders open to skilled labor migration. It might also have been able to persuade the other Arab governments to collect taxes for Egypt from its workers abroad. While the ostracism of Egypt from the Arab nation continues, the acceptance of Egyptian nationals as fellow Arabs is its only important tie to other Arab countries. However, in preserving this relationship, Egypt permits itself to be exploited.

Structural community almost always involves exploitation of the weak and a general loss of national autonomy for participating countries. Structural linkages flourish as a result of national policy. The costs of the independent, sub-governmental decisions that result in structural ties are also borne by the governments. These costs may be mitigated by government-to-government agreements, either by themselves or through international organizations. It is this kind of community to which we now turn.

Institutional Community in OAPEC

The eclectic community model developed in Chapter 1 defines institutional community as sectoral integration directed by an international regime such as a common market agreement, a set of international financial institutions, or a charter of another kind of international organization such as OAPEC. Varying degrees of institutional community may be present in different ecological communities or in the same community at different times. The end points of the community continuum were called "alliance" and "identification." An alliance is a relationship that may be so distant that cooperation or the allocation of value and resources is only potential. This is an arm's length political or economic relationship, and often may be motivated by nothing more than each partner's desire to avoid domination by the other. One or more self-centered goals may be at the root of all interstate compacts when they are first made. But in many instances this is diluted by other goals, such as regional, ethnic, or religious solidarity, or perhaps a need to submerge economic nationalism in order to survive as politically independent nations.[40] Either a desire or a need to favor joint over individual planning and action leads to a strengthening of existing structural ties and to the development of new ones.

If the benefits of an international organization are perceived as valuable and are equitably shared by all its members, a process of identification with the organization by the members occurs. Identification can be inferred from sectoral integration in the jurisdiction of the international organization. Degrees of loyalty to the international organization short of total identification may be approximated as the proportion of nationally controlled plans to joint activity controlled plans.

OAPEC is a single sector organization with no acknowledged ambitions other than to coordinate member petroleum policy, particularly in downstream investment. It aspires to organizational autonomy and to the ability to implement successfully its policy decisions. This is shown in its charter provisions, which permit OAPEC to make policy binding on all its members by majority vote rather than by consensus as OPEC and the Arab League must. Paradoxically, OAPEC authority is supported by the requirement that special projects be spun off from the parent organization. The independence of the joint venture companies reinforces members' perceptions of them as economic entities and not as political threats to national autonomy.

The OAPEC record on oil policy coordination is mixed. To assess this record in terms of the degree of institutional community evident among the members of OAPEC, two projects will be compared: the tanker company, AMPTC (the Arab Maritime Petroleum Transport Company), and the dry-dock project, ASRY (the Arab Shipbuilding and Repair Yard). In this first analysis, institutional community will be evaluated in terms of the degree of sectoral integration to be found in each of these project areas among the member states of OAPEC. Second, aggregate member commitment to OAPEC will be measured by its administrative budget totals and project capitalization. It would be helpful to be able to look at specific budget allocations as well, but this information is not available. As a part of this analysis, some judgment with regard to members' private benefits arising from project siting will be made. Finally, the ability of the organization to assert its autonomy from its members and to continue to carve out important roles for itself in the area of petroleum policy will also be discussed.

Policy Coordination and Sectoral Integration

Although OAPEC would prefer to serve as an investment coordinator for its members' downstream ventures, in practice it has also been cast as a role model. OAPEC success has bred nationalistic imitation, particularly in maritime petroleum transportation. When AMPTC was set up, no other Arab-owned tanker fleet existed. Since then, such fleets have proliferated. A direct comparison of nationally owned to OAPEC joint venture investment in tankers can be made using data from OPEC (see Tables 9 and 10).

Three-fourths of Arab public investment in tanker capacity in 1978 was under national rather than joint control. This led to competition among the Arab tanker companies during the 1974-78 period of declining demand for petroleum from the Middle East. Intra-Arab competition reduced profits for all and, perhaps even more critically, spread scarce Arab technical man-power over several installations instead of concentrating it to better effect in only one.[41] Even if OAPEC could coordinate all nationally owned and joint tanker companies in what in effect would be an Arab petroleum shipping cartel, only the profits part of the problem would be addressed. Competitive behavior in downstream investment at this early stage in the economic development of the region is counterproductive precisely because of the strains it puts on native capacity. Indeed, such wastage of scarce human resources contributes to a situation of even greater dependence on the multinational oil companies for technical services than the oil-exporting countries experienced when the companies controlled their oil.[42]

TABLE 9 *Arab Oil Tanker Fleets*

OAPEC	Nationally Owned[a]	
2,084,998 d.w.t.	Kuwait	2,312,760 d.w.t.
	Algeria	743,000 d.w.t.
	Iraq	1,396,877 d.w.t.
	Saudi Arabia	1,589,877 d.w.t.
	Egypt[b]	126,499 d.w.t.
	Libya	1,161,000 d.w.t.
	Total	7,330,013 d.w.t.

Sources: Total OPEC country capacity from OPEC, *Annual Report, 1979* (Vienna, 1979), p. 91. OAPEC tanker fleet capacity and flag assignment from OAPEC *Bulletin*, February 1977, p. 11.

[a]Does not include tonnage owned by private companies.
[b]Data from *OPEC Bulletin*, 1 August 1977, pp. 1-3.

The dry dock company stands in contrast to AMPTC as a project that has attracted more complementary (potential) imitators than competitive ones. Shortly after plans for an OAPEC dry dock were announced, Iran, Iraq, and Dubai made public their own plans for dry docks.[43] Iran and Dubai intended to compete directly with the OAPEC yard with regard to the size of the ships their own ventures were to be designed to handle.[44] Iraq alone proposed that its future facilities be designed to be complementary to OAPEC's by concentrating on smaller ships.[45] The press speculated that the announcements by

Iran and Dubai were intended as challenges to see which of the three ventures—OAPEC's, Iran's, or Dubai's—would be first with its project. Presumably, the first would be the only venture, as "the implementation of more than one such project would present unnecessary competition and tend to undermine the economies of each individual scheme."[46]

TABLE 10 *Arab Gas Carriers*

OAPEC	Nationally Owned	
132,050 c.m. (LPG)	Algeria	11,850 c.m. (LPG)
		424,369 c.m. (methane/LPG; LPG)
	Kuwait	217,754 c.m. (LPG)
	Saudi Arabia	101,902 c.m. (LPG)
	UAE	1,546 c.m. (LPG)

Sources: OAPEC *Bulletin*, February 1977, p. 11; OPEC, *Annual Report, 1979* (Vienna, 1979), p. 93.

The OAPEC dry dock project, ASRY's most important installation, was the first to break ground. Construction proceeded as scheduled and the facility began operations in late 1977 as planned. At about the same time as the official opening of the ASRY dry dock, the contract to build a nationally owned slipway repair yard (for small vessels of up to 1,000 tons) was let by Dubai to a Japanese firm.[47] In spite of its threats to compete directly with the OAPEC venture, Dubai has apparently opted for a complementary installation, at least for the present. Plans for dry docks in Iraq and Iran have not been implemented.

If these two projects had been compared in mid-1978 and evaluated on their relative contributions to sectoral integration in the Arab oil industry, one would have been judged a failure and the other a success. AMPTC lost money steadily. Part of the reason for this was the general depression in the tanker market, but part of it was the competition from the nationally owned fleets and private venture fleets owned entirely or in part by nationals of the member countries. Concerned about the profitability of these ventures, member countries refused to devise a system of preferences to benefit the OAPEC joint venture fleet. Operating losses resulted from the idleness of some of the fleet much of the time.[48]

ASRY also has lost money during its early operations, but it did so through competitive pricing.[49] The ASRY dry dock has operated at or near capacity since it was opened in 1977. More important, it operated without direct competition from other Arab installations. The ship repair industry was as depressed as the tanker industry, and for the same reasons, but ASRY managed to keep working while some AMPTC ships lay idle for long periods. This was the result of structural differences in the two industries, however, not to greater member cooperation with ASRY than with AMPTC. There was no other dry dock in the area when ASRY's facilities were planned and built; there were no other Arab tanker companies when AMPTC was founded.

The difference between the ASRY monopoly and the competitive Arab tanker market is probably related to differences in entry costs into the two industries and their differing potentials for vertical integration. While the cost of a national dry dock might be prohibitive to all but the richest countries, a single tanker does not put the same strain on a country's budget. The ASRY dry dock cost $287 million. An oil tanker can be purchased for less than a third of that amount. If an oil-exporting country chooses to buy a tanker, it can guarantee itself a minimum amount of business by shipping its own oil. In the absence of a huge national fleet, ship repairing does not lend itself to vertical integration in the same way that oil transport does. An Arab country could not guarantee itself enough business to survive, much less to prosper, as the owner of one among several regional dry docks.

That these economic reasons probably dominate the causes of the different development of the two industries can be inferred from the political situation of each one. The tanker company was founded on the premise that AMPTC would be "an integral and central part of the combined capacities of the national fleets of its shareholders."[50] Once it began operations, however, the OAPEC members, having "recently gained control of their oilfields and acquired some financial resources . . . tended to emphasize their preference for their own national tanker fleets."[51] Thus, until mid-1978, nationalism dominated the attitude of OAPEC members toward their joint venture tanker fleet.

As has been noted, ASRY's competition has remained potential. But acrimony among the membership complicated the early history of this company as well. Before the dry dock had even been designed, anger over the choice to locate it in Bahrain rather than in Dubai led Dubai, one of the Gulf Emirates, to withdraw from OAPEC. Dubai then began studies preliminary to mounting a rival effort of its own.[52] Had a Dubai project been the first to break ground, ASRY might be in the same position as AMPTC.

In 1972, when the conflict over the location of the dry dock was at its height, OAPEC was just recovering from the membership crisis that had dominated its internal politics since 1970 (see Chapter 2). Contemporary OAPEC officials chose to pretend that there was no conflict among the members. More recently, the withdrawal of Dubai has been acknowledged in official publications.[53] Since 1972, OAPEC has begun to reassert its automony, seriously weakened by policy changes that formed a part of the membership crisis resolution. Nineteen seventy-eight was a year of major breakthroughs for OAPEC. In May, protocols establishing the Judicial Board and the Training Institute, policies that had evolved painfully over a long period of time, were signed by the members of the Council of Ministers. In September, the council finally extended its authority over member government behavior in the tanker industry by passing a set of resolutions designed to rescue AMPTC from death due to capital erosion and high operating losses, at least for five years.[54] The strong position taken by member government oil ministers plus the criticism of the governments' competitive attitude by opinion leaders such as Abdullah Tariki[55] may be enough to tip the balance between national-ism and integration of Arab tanker policy. There is also the possibility that OAPEC member governments can eat their cake and have it too. The oil market as a whole tightened up considerably during the Iranian revolution, and, despite a subsequent period of oil glut in 1980, it is likely to remain generally tight in the medium to long term (see Chapter 5). Tight markets give greater leverage to the producing governments. They should be able to demand that more oil exports be transported in Arab tanker fleets. If such a move were to succeed, the oil ministers could avoid having to arrive at a redistributive policy to allocate insufficient demand for the services of Arab tankers among the national and jointly owned fleets.

In conclusion, sectoral integration in the Arab ship repair industry is quite advanced. The only major facility in the region is owned by an OAPEC joint venture company. Although there have been repeated threats from Dubai to build its own dry dock to compete with the joint venture project, these threats have been scaled down to the point that the dry dock Dubai contracted for in 1977 is planned to serve only small ships. This means that ASRY will have no Arab competitor. Although a nationally owned dry dock of any size will signify a reduction in sectoral integration in the ship repair industry, a complemen-tary facility does not threaten the future of collective action in the ship repair industry as a direct competitor would. The possibility of OAPEC-dominated "coordination" of the ship repair industry remains plausible.

In spite of a jointly owned project in the maritime transport industry, the Arab tanker industry could not be said to have been sectorally integrated at all until mid-1978.[56] Nationally owned tanker fleets competed directly with the AMPTC fleet until the passage of the 1978 resolutions, which required member governments to allocate enough business to AMPTC to keep its ships employed. As it stands, the resolutions are a statement of intent on the part of the member governments to pursue sectoral integration in the tanker industry. It is too soon to judge whether this intent can become a reality.

Financial Support of OAPEC

Another way to view member support of OAPEC is through their financial contributions to the organization. This can be done in an absolute and in a relative sense. First, as an absolute measure of support, one would want to know how much money the member governments contribute to OAPEC and whether their financial support has increased or decreased over time. Second, using sectoral integration as the model of complete support, one would wish to know the relative amount of resources allocated to industrializing ventures in the oil industry through OAPEC as compared to the resources devoted to nationally controlled projects.

Aggregate statistics for the Administrative Budget of OAPEC for its first eleven years are given in Table 11. Oil revenues received by the members from 1970 to 1975 are reported in Table 12. The patterns of change from year to year are different in these two tables. While aggregate oil revenues increased throughout the period, support for OAPEC appears to have decreased substantially from the previous year on two occasions. The first was in 1971 and the second in 1974. Since details concerning the allocation of the OAPEC budget are not released to outsiders, one may only surmise why

TABLE 11 OAPEC's Administrative Budget

Year	Allocated*	Per Member*	Spent*	Percentage Change in Allocation	Allocated Less Spent*
1968	29,974[a]	9,991[a,b]	27,154[a]	—	2,820[a]
1969	209,079	69,691[b]	137,088	—	71,991
1970	272,141	90,714[b]	212,230	30.2	59,991
1971	257,484	32,186[c]	181,856	-5.4	75,628
1972	329,851	41,231[c]	206,651	28.1	123,201
1973	482,408	48,241[d]	404,115	46.3	78,294
1974	345,582	34,558[d]	342,179	-28.4	3,403
1975	796,000	79,000[d]	758,839	130.3	37,161
1976	1,327,000	132,700[d]	n.a.	66.7	n.a.
1977	1,460,000	146,000[d]	n.a.	10.0	n.a.
1978	1,650,000	165,000[d]	n.a.	13.0	n.a.

Sources: OAPEC, *Secretary General's Second Annual Report* (Kuwait, 1975), p. 79; letter dated 29 November 1978 from Azmi Tubbeh of the OAPEC information office. The letter was the source of data from 1976 to 1978. All other data from the *Annual Report* as noted.

*Currency expressed in Kuwaiti dinars. Per member budget allocations determined assuming equal assessments—see Appendix 1, Article 27.
[a]Not a full year. Funding period ran from 14 October to 31 December 1968.
[b]There were three members during this period: Kuwait, Libya, and Saudi Arabia.
[c]Members included the three founders plus Algeria, Bahrain, Qatar, Abu Dhabi, and Dubai.
[d]Membership the same as at present: Kuwait, Saudi Arabia, Libya, Algeria, Bahrain, Qatar, the United Arab Emirates, Egypt, Syria, and Iraq. Members were counted as contributors beginning the year after their admission to OAPEC. Dubai and Abu Dhabi were independent members until Dubai withdrew from OAPEC in December 1972. After that, the Emirates were represented collectively as a single member of OAPEC.

TABLE 12 Oil Revenues of OAPEC States in Millions of U.S. Dollars

Country	1970 Revenue	1971 Revenue	1971 % Change	1972 Revenue	1972 % Change	1973 Revenue	1973 % Change	1974 Revenue	1974 % Change	1975 Revenue	1975 % Change
Algeria	325	350	7.7	700	100.0	900	28.6	3700	311.1	3375	-8.8
Bahrain	—	n.a.	—	36	—	49	36.1	262	434.7	300	14.5
Iraq	521	840	61.2	575	-31.5	1843	220.5	5700	209.3	7500	31.6
Kuwait	895	1400	56.4	1657	18.4	1900	14.7	7000	268.4	7500	7.1
Libya	1295	1766	36.4	1598	-9.5	2300	43.9	6000	160.9	5100	-15.0
Qatar	122	198	62.3	255	28.8	409	60.4	1600	291.2	1700	6.3
Saudi Arabia	1200	2149	79.1	3107	44.6	4340	39.7	22574	420.1	25676	13.7
Syria	—	n.a.	—	n.a.	—	93	—	414	345.2	600	44.9
UAE	233	431	85.0	551	27.8	900	63.3	5536	515.1	6000	8.4
TOTAL	4591	7134	55.4	8479	18.9	12734	50.2	52786	314.5	57751	9.4

Source: OAPEC, Secretary General's *Third Annual Report* (Kuwait, November 1976). Percentage changes for total figures not given in source. Data for Egypt not available.

the earlier decrease took place. The years 1970 and 1971 were turbulent for the organization because it changed from a group of politically conservative Arab oil exporters to a more inclusive organization of all Arab states earning a substantial portion of their national incomes from oil. The 1971 budget cut may have been a reflection of indecision on the part of OAPEC members about their commitment to a changing organization.

According to an official of the OAPEC Secretariat, the budget drop in 1974 was "actually a sharp increase in 1973 over 1972. The reason is that during 1973 an amount was allocated for carrying out the feasibility studies for the Dry Dock facility in Bahrain...."[57] Another reason for the drop in financial support for OAPEC in 1974 might have been related to the sudden increase that took place in that year in the number of Arab funds and development banks. Table 13 shows the Arab funds and development banks established from October 1973 to December 1974.

During and after the oil price rises, billions of dollars, primarily those earned by oil exporters, were committed to developing country assistance by other developing countries. OPEC established its own fund about this time and also contributed heavily to the International Monetary Fund's oil facility, set up to help countries having severe problems paying for their oil at the suddenly much higher rates.[58] OAPEC also contributed funds to offset the

TABLE 13 *New Arab Funds and Development Banks, 1973-1974*

Fund/Bank	Date*	Capitalization
Arab Bank for Economic Development in Africa[a]	November 1973	$ 231 million
Arab African Oil Assistance Fund[b]	January 1974	$ 200 million
Arab-African Technical Assistance Fund[a]	January 1974	$ 25 million
Iraq Fund for External Development[c]	November 1974	$ 169 million
Saudi Development Fund[d]	September 1974	$2,800 million
Islamic Development Bank[e]	August 1974	$2,400 million

Source: OAPEC, *Secretary General's Second Annual Report* (Kuwait, November 1975).

*Date of the incorporation agreement.
[a]Members: Arab League countries.
[b]Members: Algeria, Iraq, Kuwait, Libya, Oman, Qatar, Saudi Arabia, and UAE.
[c]Member: Iraq.
[d]Member: Saudi Arabia.
[e]Members: 27 Arab and Islamic countries.

burden of increased oil prices by granting interest-free loans to Arab League members "not at the same time members of OAPEC and/or substantial producers of oil."[59] The willingness of oil-exporting countries to share their new wealth helped to maintain developing country solidarity during the conferences on the creation of a "new international economic order."[60] Foreign assistance given by four Arab oil-exporting countries during 1973-76 actually exceeded 5 percent of GNP.[61] This vastly increased allocation of money to foreign aid reduced total Arab resources available for organizations such as OAPEC.

Capitalization of the OAPEC projects is shown in Table 14. Initially, shares in each corporation are divided equally among the membership. Members not wishing to take part or all of their shares may offer them back to the group, where they are again equally divided among the remaining members.[62] Shareholders participate in any profits made by the corporations in proportion to their equity shares. The types of benefits generated by OAPEC projects take the following forms: private profits distributed to investing governments; collectively shared systems of information, education, petroleum services, and representation abroad; and the redistribution of wealth

TABLE 14 *Member Equity Shares of Subscribed Capital*

Member	AMPTC % Equity	ASRY % Equity	APICORP % Equity	APSC % Equity
Saudi Arabia	13.56	18.84	17.00	14.00
Kuwait	13.56	18.84	17.00	14.00
UAE	13.56	18.84	17.00	14.00
Iraq	13.56	4.70	10.00	3.00
Qatar	13.56	18.84	10.00	10.00
Bahrain	5.00	18.84	3.00	3.00
Libya	13.56	1.10	15.00	17.00
Algeria	13.56	—	5.00	10.00
Egypt	0.10	—	3.00	5.00
Syria	—	—	3.00	10.00
Subscribed Capital in Millions of U.S. Dollars	$ 500	$ 340	$ 340	$ 51

Source: Ali A. Attiga, "Regional Cooperation in Downstream Investments: The Case of OAPEC" (paper delivered at the OPEC Seminar on the Present and Future Role of the National Oil Companies, Vienna, 10-12 October 1977), Table 11.

Note: The distribution of equity shares in the new joint venture, the Arab Engineering Consulting Company, has not been made public.

through the transfer of capital from several members to build an installation in one. OAPEC delivers a range of benefits, including redistribution, similar to those provided by national governments to their citizens.

Compulsory redistribution is a unique feature of OAPEC. The economic funds discussed above are voluntary organizations, dispensing loans and grants that impact on the benefactors as pure public goods, that is, they promote developing-country solidarity, which improves the bargaining positions of recipients and benefactors alike in dealing with developed countries and multinational corporations. OAPEC project funding is obligatory under the formal international agreement chartering each company. Project impact on all but one member per facility is redistributive of jobs, infrastructure, and prestige.

The pattern of OAPEC redistribution through siting is shown in Table 15. Although the information listed is incomplete, one aspect of the data commands immediate attention: no direct OAPEC benefits are reported to have been distributed to Syria,[63] in spite of its (obligatory) financial participation in the Arab Petroleum Investment Corporation (APICORP) (3 percent) and its (voluntary) participation in the Arab Petroleum Services Company (APSC) (10 percent). Saudi Arabia, Kuwait, and the United Arab Emirates have contributed more than the other members to the financing of OAPEC projects and receive a somewhat smaller proportion of the benefits, although each enjoys the prestige of hosting a project headquarters. Bahrain, Algeria, and Egypt have each received OAPEC facilities, including several APICORP loans for Algeria. OAPEC's treatment of Syria appears as a glaring anomaly in its redistributive record. In contrast, the dry dock siting in Bahrain, APSC assistance to oil exploration in Libya, and APICORP loans and grants to Egypt and Algeria are all examples of redistribution of resources from the better off to those who are not so well off. [64]

Member financial support of OAPEC has been very high. Including budget allocations for its first eleven years and the total capitalization of the four joint venture companies, OAPEC members have already spent in excess of $1.3 billion. The OAPEC administrative budget has increased monotonically since 1974. Beginning in 1978, although not included in available totals, are increased obligations for the new training institute (estimated "conservatively" by one source to involve capital costs of $12 million),[65] costs associated with the implementation of the Judicial Board protocol, and the distribution of AMPTC's operating losses among its nine shareholders. At the close of the 1970s, member financial support of OAPEC and its project companies was high and increasing.

The $1.3 billion total spent over eleven years by OAPEC members on OAPEC administration and projects might be compared to the foreign aid record of these states. Table 16 reports this aid as "net flows," meaning that

TABLE 15 *OAPEC Project Siting*

Project	Subscribed Capital[a]	Headquarters	Facility Assignments	Percentage of Facility Capacity
AMPTC	500	Kuwait	Qatar	6.5
			UAE	6.5
			Kuwait	6.7
			Saudi Arabia	14.2
			Iraq	33.7
			Algeria	18.5
			Libya	15.3
AMPTC (gas)			Kuwait	50.0
			Saudi Arabia	50.0
ASRY	340	Bahrain	Bahrain	100.0
APICORP[b]	340	Saudi Arabia	Egypt	n.a.
			Algeria	n.a.
			Qatar	n.a.
			Jordan[c]	n.a.
			Bahrain	n.a.
			OAPEC	n.a.
			SUMED[d]	n.a.
			Morocco	n.a.
			Kirkuk-Alexandria Pipeline	n.a.
APSC	50.7	Libya	Libya	100.0[e]
AECC	12	Abu Dhabi	n.a.	n.a.

Sources: OAPEC, "Facts About OAPEC" (Kuwait, 1976); OAPEC, *Bulletin,* November 1975-December 1978; OAPEC, *Secretary General's Fourth Annual Report* (Kuwait, 1977), pp. 85-89; *MEES*, 23 June 1980, p. 7.

[a]In millions of U.S. dollars.
[b]Siting includes loans and grants.
[c]Not a member of OAPEC.
[d]Project site is Alexandria, Egypt. Shareholders are: Egypt, 50 percent; Saudi Arabia, 15 percent; Kuwait, 15 percent; UAE, 15 percent; and Qatar, 5 percent.
[e]Includes work done through end of 1979, but does not mean that entire company was committed to subsidiaries and projects in Libya only.

OAPEC countries giving aid to other developing countries are shown as negative donors in years when their own aid receipts exceeded payout. OAPEC members collectively spent more on foreign aid in each of the four

TABLE 16 Total Net Flows[a] from OAPEC Members[b] to Developing Countries, 1973-1976

Member	1973		1974		1975		1976	
	Amount	% of GNP	Amount	% of GNP	Amount	% of GNP	Amount	% of GNP
Algeria	29.7	0.36	51.4	0.43	42.2	0.31	-66.6	-0.43
Iraq	11.1	0.21	440.2	4.16	-251.4	-1.91	-119.7	-0.75
Kuwait	550.0	9.17	1250.1	11.46	1711.2	11.44	1874.8	11.50
Libya	403.7	6.25	263.2	2.21	362.8	2.96	373.2	2.43
Qatar	93.7	15.62	217.9	10.90	366.7	16.90	245.4	10.37
Saudi Arabia	334.9	4.12	1622.1	7.19	2466.7	7.42	2826.0	7.09
UAE	288.6	12.03	749.4	9.78	1206.6	13.59	1143.8	11.45
TOTAL	1711.7		4594.3		5904.8		6276.9	

Source: Ibrahim Shihata and Robert Mabro, "The OPEC Aid Record," The OPEC Special Fund (Vienna, January 1978), Table IV-14.

[a] In millions of U.S. dollars.
[b] Data unavailable for Bahrain, Egypt, and Syria.

reported years than they did on OAPEC over eleven years. This comparison puts member support of OAPEC in a somewhat different perspective. $1.3 billion is an impressive investment for a group of developing countries, but it is less remarkable given the current account surpluses of many members of the group and their expenditures on foreign development assistance.

In addition, OAPEC is not the only avenue for member investment in the Arab petroleum industry. Each country has a national oil company, performing services for the governments similar to those operating companies perform for multinational corporations. Several member countries have extensive investments in national tanker companies. All ten member countries have installed some primary refining capacity. This reached a total capacity of 2,596,000 barrels per day at the end of 1978.[66] Some OAPEC countries, most notably Kuwait, also invest heavily in research into alternative energy sources.

It appears then that OAPEC absorbs only a fraction of its members' total disposable resources. In turn, its role in the Arab oil industry can only be described as minor when measured in terms of the resources devoted to the industry as a whole. If one can conclude that OAPEC is of major importance, it is not because of its share of the total Arab oil industry but rather because of how that share is used.

OAPEC performs its role as the Arab petroleum organization in two ways: substantively, that is, *what* it does, and procedurally, that is, *how* it does what it does. Substantively, OAPEC assists its members in coordinating their respective national petroleum policies through facilitating the exchange of information and by sponsoring research. In addition, it has made remarkable efforts to train a native workforce for all levels of the oil industry. Finally, it has sponsored joint, cooperative oil investment, particularly valuable to member governments with low levels of oil reserves and/or low levels of per capita income.

Procedurally, these functions have been categorized as passive or active implementation of the organization's objectives. Passive implementation corresponds to the more traditional approach taken by the Arab League and by OPEC in collective oil policymaking. Participation is optional. For example, member governments make available whatever information they individually choose for the *Annual Statistical Report*. Over time the range of information supplied by this periodical has broadened, but, cross-nationally, some data is still incomplete.[67]

Active implementation is different because it is compulsory—at least, it is as compulsory as formal international agreements can ever be. The OAPEC joint venture companies are created by international agreement. They are then spun off from the parent organization. This process liberates them from OAPEC as they are liberated by charter from member governments. Liberation is not abandonment, however. Member governments have been required to furnish substantial financial support to ailing OAPEC companies. In return, every OAPEC company, regardless of its financial health, has acknowledged its obligation to contribute to the formation of a trained, native, petroleum industry workforce. The OAPEC member governments have created a novel corporate structure in the joint venture companies to deal with the problem of industrializing while decreasing Arab dependency on developed countries. Nationalistic investment does not have as good a record in this regard.[68] Yet, it is hoped that the companies will eventually be competitive with private investment ventures. OAPEC has tried to deal with the problems of economic inefficiency generally associated with government-owned industries by granting the companies substantial independence and by exhorting them to make profits. There is naturally a contradiction between profitmaking and social activism. It will be interesting to see if the companies can manage both when market conditions improve.

A second measure of OAPEC activism is the new training institute in Baghdad. The training institute is an ambitious project that seeks to go beyond technical training as a company cottage industry and make it a direct concern of the organization. By establishing an institution to train instructors, OAPEC is creating a support service for the companies that have had to send many of their trainees abroad for instruction. The training institute was set up with an eye to its potential as a powerful integrative force in the region's oil industry and thus as an influence for rationalizing member petroleum industry practices. If the training institute is able to function as its founders hope, it would also be an excellent vehicle for the transfer of technology and, indeed, an instrument suitable for building a native Arab technology.

Finally, the Judical Board is an important advance in institutional capability, especially as compared with efforts by the Arab League and OPEC. Each of these organizations has also considered setting up a court, and, like OAPEC before May 1978, each has been unable to agree on one. Lynn Mytelka has pointed out that cooperative international structures have a better chance of succeeding if procedures and obligations are worked out beforehand and if participants have some flexibility in adjusting to them.[69] Because of the harmony of interests among OAPEC's founding members, the jurisdiction of the Judical Board and member obligations under it were made a part of the original OAPEC Agreement in 1968. The joint venture companies and the Judical Board are indications of the changing balance between national and supranational authority in OAPEC. From its inception, OAPEC was designed to be an institutionally strong organization. It began creating novel, powerful international institutions with the joint venture companies. Because participation in the joint venture companies is not compulsory, members have some flexibility in their initial commitment to a strong, supranational authority. According to Karen Mingst, "Because of the flexibility of both participants and regulations, critical oil-related tasks have become centralized through OAPEC quite rapidly."[70] The ratification of the Judical Board protocol would seem to confirm this earlier assessment.

Conclusions

What evidence there is of Arab structural community points to most OAPEC members of the Arab League as representative of Arab core characteristics and values. This coincidence has meant that oil has not split the Arab states into oil haves and have-nots. Instead, peripheral countries have been motivated to maintain their ties to the oil states, such as Saudi Arabia, which in themselves define the ethnic and religious characteristics of an Arab identity. The inability, and for some, the disinclination of Arab oil exporters to disassociate themselves from their non-oil brethren is illustrated by the ill-fated attempt to make OAPEC an independent bloc within the Arab League. It succeeded during the very brief period when OAPEC's few founding members shared political and economic goals. Once political solidarity was broken, OAPEC's bloc behavior was quickly modified.

Interdependence in the Arab world is the result of the movement of goods, money, and people. Trade and labor migration create extensive ties among many Arab countries. These ties link oil-exporting with oil-importing countries. In each of the areas examined here, structural community among Arab countries was the result of decisions made by individuals and governments on the basis of perceived self-interest rather than of bloc behavior. Thus, structural community does not appear to isolate the Arab oil exporters from the rest of the Arab League. Whether first- or second-level effects are considered, one can only conclude that oil wealth as such has not shown itself to be a cause of divisiveness.

Institutional community within OAPEC has gone through two transitions since the organization was founded. Originally, institutional community was very high. The OAPEC charter provided for strong central procedures and institutions, reflecting the high degree of similarity in the political and economic viewpoints of the founding members with regard to oil policy. The Libyan revolution destroyed this similarity, and centralization gave way to increasing nationalism. This eventually showed itself in the relationship between member governments and the joint venture companies, as national enterprises that competed directly with the OAPEC tanker company were created. But in the fall of 1977, OAPEC officials began to assert publicly the need for the type of oil policy coordination OAPEC was set up to provide and which member nationalism was destroying. In 1978, the Council of Ministers took three actions signifying a reversal of the nationalistic trend. It agreed to set up the OAPEC Judicial Board, provided for in the original OAPEC

Agreement; it required the shareholders in the tanker company to provide financial assistance and cargo preferences to the OAPEC fleet; and it allocated funds to cover operating losses by the dry dock company.

Sectoral integration of the ship repair industry through OAPEC had always been high; the new policies marked a move toward some sectoral integration in marine transport where virtually none had existed. The policy establishing the OAPEC court goes beyond sectoral integration and suggests that the Judicial Board could become an entirely new international regime in the oil policy area.

The signing of the Egyptian-Israeli peace treaty in March 1979 precipitated a crisis for institutional community in OAPEC, the Arab League, and other Arab organizations. The OAPEC regime and others like it proved to be weaker than the anti-Israel residue of Pan-Arabism, thought by one observer to have exhausted its power to dominate politics in the region.[71] Instead, communal enmity toward Israel disrupted institutional ties between Egypt and other Arab countries. Strangely enough, rational, bureaucratic organizational community, to use Max Weber's term, has proven to be more fragile than structural community based on individual behavior. This means that Ajami may be correct in his view that Pan-Arabism is a dying force in the Middle East, but for the masses rather than for their governments.

OAPEC is not itself a structural community. For the most part, it is an institutional community whose power and autonomy are increasing. The behavior of the other Arab governments after the Egyptian-Israeli treaty signing shows that institutional community is independent of structural ties and can be broken much more easily. Egypt's long-run position in OAPEC will be an important indicator of the relative power of economic and communal ties.

Notes

1. See Fouad Ajami, "The End of Pan-Arabism," *Foreign Affairs* 57 (Winter 1978/79).

2. The text of Decisions 9 and 10 of the Baghdad Summit Conference held in November 1978, were published in *Al-Safir* (Beirut) on 21 March 1979. The translated text was published in *MEES*, 26 March 1979, pp. 4-5, and includes provision for moving the headquarters of the Arab League out of Cairo and the outline of the Arab boycott of Egypt. The quotation in the text is from page 5.

3. Walter J. Levy, "The Years That the Locust Hath Eaten: Oil Policy and OPEC Development Prospects," *Foreign Affairs* 57 (Winter 1978/79), discusses conflict within the region as interfering with a rational development policy. Possibly the most serious regional conflicts have broken out between Egypt and Libya, and this in spite of many structural and institutional ties between the two countries.

4. Karl Kaiser, "The Interaction of Regional Subsystems," *World Politics* 21 (October 1968).

5. Richard Rosecrance, Alan Alexandroff, Wallace Koehler, John Kroll, Shlomit Lacquer, and John Stocker, "Whither Interdependence?" *International Organization* 31 (Summer 1977).

6. Ibid.

7. Robert Keohane and Joseph Nye, *Power and Interdependence* (Boston, 1977), Chap. 1.

8. See Mary Ann Tetreault, "Measuring Interdependence," *International Organization* 34 (Summer 1980).

9. Albert H. Hourani, *Minorities in the Arab World* (London, 1947), p. 1.

10. The groups are OAPEC: Algeria, Bahrain, Egypt, Iraq, Kuwait, Libya, Qatar, Saudi Arabia, Syria, and the UAE; Other Arab League: Jordan, Lebanon, Mauritania, Morocco, Oman, Somalia, Sudan, Tunisia, the Arab Republic of Yemen (ARY-North Yemen), the People's Democratic Republic of Yemen (PDRY-South Yemen). Non-oil Arab countries were chosen in order of their joining the Arab League on the assumption that the newer Arab League members are ethnically and geographically progressively farther from the "core" of the Arab World. Albert Hourani, *Minorities in the Arab World*.

11. Christopher S. Wren, "Peace Hasn't Been Easy for Egypt's Army," *New York Times*, 23 September 1979, p. 2E.

12. Anthony Sampson, *The Arms Bazaar* (New York, 1977), Chaps. 1, 9, 14, 18.

13. Oil-Rich Kuwait, Saudi Arabia, and Libya have spent heavily on the military needs of other Arab states since 1967.

14. Based on 1975 data from the United Nations, *Yearbook of International Trade Statistics 1976* (United Nations, 1977).

15. Barry W. Poulson and Myles Wallace, "Regional Integration in the Middle East: The Evidence for Trade and Capital Flows," *The Middle East Journal* 33 (Autumn 1979).

16. For example, on the Yanbu project alone the following contractors were used: Ralph M. Parsons, Dames & Moore—Basil Geotechnics, Furgro, Inc., Geotek Alireza, Abdul Rahman Al-Namlah Establishment, H. B. Zachry, Eppco Marine, Sassakura Engineering Co., Nippon Electric Co., and Projects Execution Establishment (reported in *Middle East Economic Survey* [*MEES*], "Supplement," 3 April 1978). The same document also lists the contractors for the older Jubail project, another international list. In addition, the U.S. Army Corps of Engineers is involved in several projects relating both to military installation upgrading and general construction assistance (see *MEES*, 29 May 1978, pp. i-iii).

17. Recycling also involves depositing oil money in lending institutions that channel it to oil-importing countries as a way to finance these imports.

18. Sampson, *The Arms Bazaar*.

19. The study was begun in Beirut in 1969 by Eugene Makhlouf, Antoine Zahlan, and Elias Eid. Eid is an engineer. Zahlan and Makhlouf write about science policy in the region.

20. Nabeel Shaath, "High Level Palestinian Manpower," *Journal of Palestine Studies* 1 (Winter 1972), as reprinted by the Arab Information Office (New York, n.d.), p. 13.

21. Ibid., p. 14.

22. J. S. Birks and C. A. Sinclair, "Egypt: A Frustrated Labor Exporter?" *The Middle East Journal* 33 (Summer 1979), 290.

23. Ibrahim Sa'ad Eddin, "The Negative Effects of Differences of Income Among the Arab Countries on Development in Countries with Low Per Capita Income: The Case of Egypt," in OAPEC, *Sources and Problems of Arab Development* (Kuwait, 1980), p. 12.

24. Birks and Sinclair, "Egypt: A Frustrated Labor Exporter?" pp. 297-301; Eddin, "Negative Effects of Differences of Income Among the Arab Countries on Development in Countries with Low Per Capita Income," p. 11.

25. Birks and Sinclair, "Egypt: A Frustrated Labor Exporter?" p. 299.

26. Ibid., p. 300.

27. Ibid., p. 301. Eddin believes an accurate estimate is impossible ("Negative Effects of Differences of Income Among the Arab Countries on Development in Countries with Low Per Capita Income," p. 12).

28. Birks and Sinclair, "Egypt: A Frustrated Labor Exporter?" p. 290.

29. Ibid., p. 301.

30. Helen Sonenshine, "Economic Implications of the Israeli-Egyptian Peace Treaty" mimeographed (n.p., n.d.), p. 13.

31. Eddin, "Negative Effects of Differences of Income Among the Arab Countries on Development in Countries with Low Per Capita Income," pp. 24-25.

32. Birks and Sinclair, "Egypt: A Frustrated Labor Exporter?" p. 296.

33. See the text of the Baghdad Resolutions (note 2). Also interview by the author with Jamil Al-Hassani, counselor, The Embassy of the State of Kuwait (Washington, D.C., 3 July 1980).

34. Birks and Sinclair, "Egypt: A Frustrated Labor Exporter?" p. 15.

35. Eddin, "Negative Effects of Differences of Income Among the Arab Countries on Development in Countries with Low Per Capita Income," p. 16.

36. Ibid., p. 17. This problem is also discussed by Birks and Sinclair, who quote from the Ministry of Planning, *Five Year Plan, 1978-1982,* Vol. 1, p. 33: "... [G]rowing numbers of Egyptians work abroad for very high wages, if compared with domestic salaries. These individuals return to Egypt possessed of high purchasing powers, which they individually direct not to savings and investment, but to flagrant and luxurious consumption." Quote appears on p. 302.

37. Birks and Sinclair, "Egypt: A Frustrated Labor Exporter?" p. 302-3.

38. Eddin, "Negative Effects of Differences of Income Among the Arab Countries on Development in Countries with Low Per Capita Income," p. 26.

39. William A. Axline, "Underdevelopment, Dependency and Integration: The Politics of Regionalism in the Third World," *International Organization* 31 (Winter 1977).

40. A major reason for forming the European Economic Community was to create a large internal market for member industries as a way to balance the advantages of the United States in production scale economies.

41. An additional complication in the tanker situation is the increasing number of privately owned joint tanker ventures that have been set up between Saudi Arabian citizens, often members of the royal family, and foreign corporate partners. As of February 1978, seven such ventures were operating (see *MEES*, 20 February 1978, pp. 6-7). These private corporations compete with the Saudi state-owned tanker

company, Petroship, as well as with the OAPEC joint venture company, AMPTC. Because Saudi Arabia has a huge current account surplus, profits from the tanker companies in which the state holds an interest might be relatively unimportant to the government. But the private ventures and the nationally owned tanker company are direct challenges by the richest Arab oil exporter to the cooperative approach represented by OAPEC.

42. *Wall Street Journal,* 17 October 1978, p. 1.

43. *MEES,* 12 May 1972, pp. 4-5; *MEES,* 4 August 1972, p. 4.

44. *MEES,* 4 August 1972, p. 4.

45. *MEES,* 12 May 1972, pp. 4-5.

46. *MEES,* 4 August 1972, p. 4.

47. *MEES,* 10 October 1977, p. 6.

48. According to its annual report released in June 1978, AMPTC lost $4.4 million in its 1977 operations, excluding overhead and depreciation (*MEES,* 5 June 1978, pp. 3-4.

49. Exact figures are not available, but ASRY's chairman, Sheikh Khalifa bin Salman bin Mohammed Al-Khalifa, admitted to operating losses in an interview with the OAPEC *Bulletin,* December 1978, pp. 1-2. In December 1978, the seven share-holding countries agreed to subsidize anticipated ASRY losses for 1978-84 to the amount of $146 million, to be contributed according to shareholdings (*MEES,* 18 December 1978, p. 2).

50. Ali A. Attiga, "Regional Cooperation in Downstream Investments—The Case of OAPEC" (paper presented at the OPEC Seminar on the Present and Future Role of the National Oil Companies, Vienna, 10-12 October 1977), p. 11.

51. Ibid., p. 12.

52. The controversy over the dry dock location is recounted in several issues of *MEES* published in 1972. These issues are: 12 May, pp. 4-5; 7 July, pp. 4-5; 4 August, p. 4; 24 November, p. 2; and 29 December, p. 7.

53. For a contemporary example, see the interview of OAPEC Secretary General Suhail al-Sa'dawi, originally published in *al-Madinah* (Saudi Arabia, 25 June 1972) and reprinted in *MEES,* 7 July 1972, p. ii. An example of a publication: "Basic Facts About OAPEC" (Kuwait, 1976). This acknowledges that Dubai was once an independent member of OAPEC and is now represented through the UAE.

54. The resolutions included provisions to (1) require the shareholding government to fund operating losses for five years in proportion to their shares in the company; (2) require member governments to take measures to ensure full employment for AMPTC vessels at competitive rates; and (3) require members to pay their outstanding balances of subscribed capital (see *MEES,* 25 September 1978, p. 12).

55. Interview by author, 25 April 1978, in Houston, Texas.

56. Karen Mingst, "Regional Sectorial Economic Integration: The Case of OAPEC," *Journal of Common Market Studies* 16 (December 1977): 112.

57. Letter from OAPEC Information Office official, Azmi Tubbeh, to the author, dated 29 November 1978.

58. OPEC nations had established a "foreign aid record" before the oil revolution of 1973, but their contributions in loans and grants increased substantially after oil prices were increased. For an excellent discussion of OPEC aid, see Ibrahim F. I. Shihata and Robert Mabro, *The OPEC Aid Record,* OPEC Special Fund (Vienna, January 1978).

59. See *MEES,* 7 June 1974, pp. 2-3.

60. Zuhayr Mikdashi, "Cooperation Among Oil Exporting Countries with a Special Reference to Arab Countries: A Political Economy Analysis," *International Organization* 28 (Winter 1974). Mikdashi attributes OPEC's success (beginning in 1964 with the first royalty expensing agreement) to solidarity among Arab countries and among these countries and other developing nations. OPEC was represented at several of the CEIC meetings by Usamah Jamali (OAPEC, *Bulletin*, May 1976, p. 3).

61. Shihata and Mabro, "The OPEC Aid Record," p. 16. These countries were: Kuwait, Qatar, Saudi Arabia, and UAE.

62. A. Kesmat El-Geddawy, "Arab Companies Established by OAPEC," in OAPEC, *Petroleum and Arab Economic Development* (Kuwait, 1978), p. 155. The APICORP charter agreement limits maximum participation to 20 percent and the minimum to 3 percent. It is the only company to require total membership participation and to prescribe such limits.

63. APICORP loans to the Kirkuk-Alexandretta pipeline (attributed to Iraq in the table) might be viewed as an indirect benefit to Syria. An official of another Arab government told me, in confidence, that the Syrian government had demanded financing from several Arab aid-distributing organizations (including OAPEC) but without detailed proposals for projects. In the face of desirable project proposals from other countries, Syria's requests were not often fulfilled. Another reason why Syria might have been slighted could be the religious differences between its leaders and those of other OAPEC countries.

64. Libya has money, but it is a needy country in the area of technical expertise.

65. *MEES*, 22 May 1978, p. 4.

66. OAPEC, *Secretary General's Fifth Annual Report* (Kuwait, 1979), pp. 42-43. At that time, an additional 1,861,000 million barrels per day of capacity was under construction and 1,400,000 MBD over that in the planning stages.

67. OAPEC, *Fifth Annual Statistical Report, 1976-1977* (Kuwait, 1978). This is the most recent issue available in English. Balance of trade statistics are incomplete. More critically, manpower distribution statistics and information on local energy consumption are also incomplete cross-nationally.

68. *Wall Street Journal*, 17 October 1978, p. 1; Walter J. Levy, "The Years That the Locust Hath Eaten."

69. Lynn K. Mytelka, "Fiscal Politics and Regional Redistribution," *Journal of Conflict Resolution* 19 (March 1975).

70. Mingst, "Regional Sectorial Integration," p. 112.

71. Ajami, "The End of Pan-Arabism."

Chapter 5

The Role of the Arab Nations in OPEC Bargaining

The OPEC Cartel

Since October 1973, the Organization of Petroleum Exporting Countries has set the price of petroleum sold in the international market. Because of this, OPEC is often called, in academic and popular literature, a "petroleum cartel." A cartel is a system of overt coordination or collusion among several economic actors by which they try to secure monopolistic prices or profits from marketing a good that they all produce.[1] OPEC and similar organizations such as CIPEC and the International Bauxite Association (IBA) are more complex than enterprise cartels because their members are governments, or else "governments enforce [their decisions] in the pursuit of national economic objectives."[2] This in turn means that government cartel behavior is influenced by the need to achieve political objectives as well as economic goals. The ability to achieve any goal is limited by the nature of the goods that these cartels produce, namely, exhaustible natural resources. Government members must design price and production schedules that take into account ultimate supply constraints and current and future expenditure needs.[3] These limiting factors vary across OPEC members and constitute an obstacle to an identity of interests among them.[4]

Another obstacle to perfect coordination is the arbitrariness of any one market price for oil.[5] The single OPEC price refers to that for a "marker" crude; the quality and location of any particular crude determines its actual selling price, with the marker crude acting as a reference. In spite of this apparent flexibility in the pricing of individual crudes, the price arrived at by applying quality and transportation differentials to the marker crude price still amounts to a compromise that may not coincide with the preferred price of any OPEC member. It may be that no member actually is committed by reason of an assessment of its own needs to any OPEC price at any given time. In a time of oil glut, this lack of commitment might contribute to the willingness of dissatisfied governments to shave prices to get more business, and thus weaken the cartel and the price structure it has created.

A final aspect of the world oil market has led some scholars to predict that an outside actor will be responsible for the eventual collapse of OPEC. Such an actor might be part of the "exporter fringe," the major oil exporters in the world market that do not belong to OPEC.[6] Norway, Mexico, the Soviet

Union, Canada, and others might at any time decide to lower oil prices to force OPEC to back down from its pricing decisions. As we shall see below, the exporter fringe cooperates with decisions affecting world oil prices. Countries in the exporter fringe prefer to behave as "free riders," collecting the OPEC-set price and an additional differential reflecting their political reliability in case of another oil embargo.

Indicators of Cartel Success

Obstacles to coordination have not prevented OPEC from behaving like a cartel most of the time since 1973. The obvious indicator of OPEC's successful cartel behavior is its ability to maintain the world price of crude oil several hundred percent above the cost of producing it.[7] A standard for cartel success can be as modest as Paul A. MacAvoy's requirement that participants' total profits be higher than they would have been under competition[8] or as tough as Paul Leo Eckbo's definition requiring participants to maintain market prices at least 200 percent above competitive prices.[9] By the most demanding standard, OPEC would qualify as a successful cartel.

In addition, OPEC has a formal agreement pledging it to work to secure a steady income to oil-producing nations[10] and a mechanism by which to arrive at decisions binding on all OPEC members, the OPEC Conference, which meets at least twice each year to decide on oil prices and other oil policies for all OPEC member states. Other OPEC organs, principally the Economic Commission, have served as arenas in which members can confront cheating by bringing charges of disloyal behavior to the attention of all members.[11] Aggregating and focusing the disapproval of noncheaters is an informal control on individual members and acts to inhibit undesirable behavior.

Cartel Imperfections

Like many other cartels, both enterprise and governmental, OPEC is not contractually complete.[12] In order to be contractually complete a cartel must produce at the lowest possible cost, sell at its agreed-upon price, and then pool the profits and divide them according to some prearranged plan.[13] In the case of OPEC, contractual completeness has not even been approximated by a plan for production prorationing, a stated objective of the Conference since its first meeting, because members cannot agree on a basis for allocating production shares.[14] Aside from the efficiency argument for a contractually complete cartel—that the joint profit maximizing output should be produced at the least possible cost—some sort of agreement on shares, either of production or profits, might at first appear to contribute to cartel stability because it would reduce opportunities to cheat by eliminating misunderstanding as an excuse for overproduction. However, it would also reduce the autonomy of individual members, particularly the high cost producers.[15] Since autonomy is a prerogative of sovereignty, contractual completeness is even less attractive to government cartels than to enterprise cartels.

In spite of the difficulty of arriving at a prorationing policy, however, the issue is still alive in OPEC. The Long Term Strategy Committee was set up in 1978 to consider prorationing and related policy issues, including long-term pricing. The chair of this committee is Ahmed Zaki Yamani, minister of petroleum and minerals for Saudi Arabia, a country that must agree to any prorationing scheme if it is to have a chance of success. The committee made its report to the Fifty-sixth (Extraordinary) Meeting of the Conference, held in Taif in May 1980. The Conference agreed with much of the report, but, because it was based on the assumption that the oil market would remain tight, it was later decided to reconsider the entire set of proposals, including those for expanding OPEC foreign aid, because of the failure of the proposals to allow for glut conditions.[16]

Previous negotiations on prorationing led to the First and Second Production programs, developed for the 1965-66 and 1966-67 production years (running from July to July.) During the First Production Program, the amount by which OPEC's staff had estimated that world demand would increase turned out to be greater than the actual increment in total demand, and instead of each country keeping to its alloted share of the increment, Iran, Libya, and Saudi Arabia produced proportionately more and the others, less.[17] No serious attempt was made to implement the Second Production Program, several members going so far as to withdraw their pledges of support for it.[18]

Deciding on a formula for prorationing production is very hard to do. However, the reappearance of growing glut in the world oil market in the spring of 1979 made the members more interested than they have been in some time to take another look at prorationing. But, even if some formula is agreed to within the next year or two, there is no assurance that it will be more effective at coordinating member production than the informal system currently operating.

Over time, OPEC has evolved an informal method for coordinating member oil policy. The Conference makes great efforts to develop consensus positions whenever possible, with the understanding that individual members may presume that they will receive post hoc support for stronger, independent positions that do not conflict with basic joint policy.[19] This has allowed individual OPEC members wide latitude in their behavior while permitting the rest of the membership to decide later whether to support a pioneering member. The absence of a definite production-sharing plan for OPEC members allows minor cheating without forcing the membership to confront a deliberate violation of a binding agreement. It also permits a certain fluidity in leadership roles with regard to production and permits individual members to decide on conservation measures or small production increases without requiring the advice or consent of the whole membership.

A permissive attitude toward the individual behavior of cartel members is particularly crucial when a cartel is composed of governments. Governments must deal with the domestic consequences of international or transnational policies. This can limit their freedom to behave as single-minded actors in pursuit of national security as their single goal.[20] Economic policies are particularly vulnerable to domestic political pressure because of their immediate impact on citizens' perceptions of satisfaction with their governments.[21] Petroleum policy, heavily economic in its impact, is likely to involve conflict among the domestic and the international goals of a given OPEC member nation. In addition, the transnational relationships characteristic of trade relations tend to erode the autonomy of nation-states.[22] Multiple and conflicting pressures on each individual state favor the survival of flexible international arrangements that permit tolerance of occasional aberrant behavior.[23] In such situations as OPEC finds itself, discrepant stands on policy issues do not necessarily indicate organizational weakness; they may simply reflect differential responses to cross-pressuring within OPEC's membership.

Reasons for OPEC's Success

Three major factors account for OPEC's post-1973 political and economic success. One is the momentum generated by Libya's dazzling victory over the oil companies in 1970. Since this Machiavellian power play, all OPEC members have assumed greater control over oil pricing and production. As a direct result of the transfer of power from the oil companies to OPEC and because of the income effects of huge oil price increases, the nature and locus of decisionmaking on petroleum pricing has been altered. The second factor is the compatible dissimilarity of economic and political goals of OPEC members. This has permitted members to exercise significant autonomy in ways that have actually furthered collective cartel objectives. The third factor that helps OPEC to support its oil price decisions is the behavior of the exporter fringe and U.S.- and European-based multinational oil corporations. Each of these has supported OPEC by matching or exceeding OPEC-set oil prices whenever and wherever possible. Thus, there is no low-price seller in the world oil market outside OPEC to depress the general oil price level and thereby break the cartel.[24]

The reinforcing effect of successive victories on the behavior of OPEC members was foreseen by Bunker Hunt executive George Schuler, who played Cassandra in the price negotiations with Libya in the early 1970s. Because of an unforeseen shortage of tanker capacity in 1970, Libyan oil became more valuable to companies shipping to customers in Europe. The

Libyans then demanded a price increase reflecting the transportation advantage of Libyan oil. Schuler predicted that allowing the Libyans higher prices for their oil would be a virtual invitation to the Arabian Gulf exporters to demand price increases also, ratcheting oil prices upward, first at Mediterranean terminals and then at Gulf ports. Schuler envisioned this price leapfrogging as only the beginning of a new petroleum regime that would see the increase of producing country power at the expense of the power of the oil companies. He predicted that the oil-exporting countries would continually insist on renegotiating ongoing contracts whenever they perceived that they held an advantage.[25]

The Tripoli and Teheran agreements did not end oil-exporting country demands for more money and control. As Schuler predicted, demands for higher taxes on the companies were soon made by OPEC members. The demands were stimulated by substantial increases in oil company profit margins since the agreements had taken effect and by changes in market conditions that weakened the ability of the companies to resist producer demands.[26] In addition, past success in raising per-barrel prices acted to bolster subsequent efforts to increase them even more by allowing negotiating governments to sustain production cutbacks while maintaining national income at acceptable levels. Thus, even before the Arab oil embargo, in September 1973, Algeria was able unilaterally to raise its posted price to five dollars per barrel.

Conditions supporting the continuing efforts of oil-exporting governments to increase their incomes from oil were intensified by the embargo. Arab production cutbacks magnified the effect of an already tight supply situation on prices in the world petroleum market. Prices rose about fourfold during the embargo period. The price rises enabled Arab participants in the production cutbacks to enjoy an increase in their incomes from oil despite the drop in production. The longer high prices were in effect, the more strength accrued to OPEC members because they could still maintain their income levels even though they produced less oil. Some members even managed to produce current account and budget surpluses, further decreasing the level of aggregate oil production OPEC members would have to sustain in order to meet income needs.

Pre-embargo victories had already eroded the relative power of the oil companies in their negotiations with the producing countries. During the October War, the initiative passed from the companies to the producing countries in the matter of pricing and, to a great extent, in production as well. The embargo then acted as a catalyst to complete the devolution of control over oil from the companies to the producing governments.[27]

The second major contributor to OPEC success has been the group's ability to combine disparate individual national petroleum policies into an

organic structure that looks like a single collective OPEC policy to consumers and oil companies. The embargo period offers the clearest illustration
of this. The production cutbacks and oil embargo were *Arab* policies, decided
upon by members of the Organization of Arab Petroleum Exporting Countries at their 17 October meeting and implemented (to varying degrees) only
by OAPEC members. The oil price increases were OPEC policies, set by
OPEC members at their meetings throughout the fall and winter of 1973-74
and implemented by OPEC members, some of which were also cutting back
production and embargoing supplies to certain countries because of commitments to the Arab organization. Although all the producing nations were
interested in increasing their per-barrel incomes, their success in setting and
maintaining the spectacular price increases that were instituted during the
winter of 1973-74 was predicated on a supply cutback. Neither the willingness
nor the ability to absorb significant production cutbacks was equally widely
distributed across OPEC members.[28] Cutbacks for the sake of cartel pricing
policy were unlikely anyway. The lack of a production-sharing agreement that
could at the least have distributed feared losses arising from a cutback policy
more or less equitably, shows the mutual distrust of many OPEC members
and their history of competition for production shares.

In the absence of a coordinated production policy for all of OPEC, individual members have set production levels in pursuit of political objectives
independent of OPEC goals. The 1973-74 cutbacks and embargo represented the fourth attempt by Arab states to use "the oil weapon" in their
long-standing conflict with Israel.[29] The confluence of cartel pricing objectives, Arab political objectives, and petroleum market conditions made the
double policy initiative appear to be a single assault directed against oil
consumers. That it was not can be seen in the persisting intra-OPEC disagreement over pricing policy that began in December 1973. But, for the most
part, these internal disagreements did not keep OPEC members from enjoying substantial individual gains through mutually reinforcing, if not similar,
petroleum policies.

A second illustration of compatible dissimilarity in OPEC goals arose
during the post-embargo recession. Some members suspected Iraq of price
shaving in an effort to increase its market share and national income. At the
same time, Kuwait and Venezuela were individually debating whether to cut
production to conserve their oil,[30] while Libya continued to follow its own
policy, requiring production cutbacks as a means of disciplining oil companies operating on its territory. Although the Iraqi action may have contributed to the softness in oil prices, the price structure itself did not collapse[31] as
M. F. Adelman, among others, predicted that it might, because of individually
motivated production restraint elsewhere in the system.[32]

The third support for OPEC's oil price increases comes from outside the
organization. It is the willingness of other oil exporters to take the OPEC

price as a floor for their own oil prices. OPEC's success at raising oil prices in 1973 meant that the non-OPEC producers also received much higher prices for their oil.

In the short term, oil exporters in OECD are cross-pressured by their desire to make as much money as they can from their oil, while hoping for stability in international financial markets and in their relations with oil-importing developed countries. High oil prices interfere with stability, contributing to worldwide inflation and to economic insecurity in oil-importing countries.[33] In the long term, however, the policy preferences of OPEC and non-OPEC oil exporters coincide. Neither group wants to see consumer demand for oil increase while reserves are drawn down. David Howell, minister of energy for Great Britain, recently described OPEC as "a positive and stabilizing force in world oil markets," one that, given the chance, Britain would work with.[34] Even the Bank of England has advocated price stability rather than a fall in the real price of oil to encourage efficiency and to reduce dependence on oil for energy.[35] Norway and Mexico both find it more rational from their perspectives to produce only enough oil to satisfy demands for income and domestic investment, thus limiting non-OPEC production and supporting high prices.[36]

The multinational oil companies are as interested as the non-OPEC oil exporters in windfall profits and have made no discernible move to push prices down. The Saudis believe that during the height of the Iranian crisis, oil companies were buying their oil at the lowest price in OPEC and then selling it in the spot market for over forty dollars per barrel, an action that if true, contributed to higher official OPEC prices in 1979 and 1980. Oil company profits were at an all-time high in 1979, the year of the Iranian crisis and the spot market dominance of news on oil prices.[37] Because pricing in large companies is rarely, if ever, a function of marginal costs but rather a percentage of gross receipts or some other shorthand formula,[38] it is in the interests of the oil companies as profit maximizers to increase their gross revenues. Adelman has called the oil companies tax collectors for OPEC. While this may be true, their behavior is also related to their own economic self-interest.[39]

These three factors, momentum, the compatibility of OPEC-member domestic policy with the aims of the organization as a whole, and the decision of non-OPEC oil exporters to become free riders on the OPEC price, account for much of OPEC's success as a cartel. Efforts of oil-consuming countries, individually or through the International Energy Agency (IEA) have been unable to lower the price of oil in nominal terms. The IEA particularly is seen by OPEC members as a direct threat to their sovereignty and to their autonomy regarding their natural resources. The IEA policy requiring member countries to create stockpiles of oil as a hedge against future oil embargoes has involved Saudi Arabia in a conflict with the United States over their desire to comply by filling their reserve with Saudi oil. On the other hand, oil consumers have made it very clear to OPEC and to other oil exporters that they are willing—and able—to pay prices even higher than the 1980 official OPEC maximum of $37.00 per barrel. This has pushed the OPEC floor price to its present level.

OPEC Bargaining Arenas

The effect of producing country policies in 1973-74 was to alter the arena of price bargaining so that all significant actors were members of OPEC.[40] Even more important, the nature of the bargaining process changed from OPEC concentration on the redistribution of power and money from the oil companies to the producing governments, to a distributive mode in which gains to the relevant actors were possible through the exploitation of outsiders without requiring redistribution of income or control within OPEC. The distributive aspects of the OPEC arena are supported by the possibility of simultaneous achievement of individual national goals of OPEC members such as oil income, production levels, conservation policy, and the use of oil as a political weapon. For the most part, OPEC members can reach their goals at the expense of outsiders rather than at the expense of other cartel members. Bargaining within OPEC has been successful in maintaining cartel unity and high price levels because it usually takes place in a distributive arena.

The concept of political "arena" used here is an expansion of theoretical work by Theodore J. Lowi and James Q. Wilson on the characteristic political environment and actor relationships associated with the formation of different types of public policy. Lowi believes that "policies determine politics."[41] By this he means that policy content determines the political arena, that is, the constellation of policymakers, interest groups, policy targets, and agencies competing for implementing authority, in which that policy is made. In other words, content determines context.[42]

Wilson quarrels with Lowi's policy parameters and categories, preferring to type policies according to the joint distribution of their associated costs and benefits.[43] Although he agrees with Lowi's basic thesis that policies determine politics, Wilson implies that a reciprocal relationship may obtain in instances where policies representing significant departures from the past are designed. Schematically, Lowi says:

$$\text{Policies} \longrightarrow \text{Politics}$$

whereas Wilson says:

$$\text{Politics} \longrightarrow \text{Policies}$$

but novel results arise when

The Wilson reciprocal causation model[44] has been chosen as a means to analyze the politics of OPEC policymaking.

From OPEC's founding in 1960 until the winter of 1973-74, the prices members received for their oil represented the outcome of conflict between the oil companies and the OPEC nations. This conflict was redistributive in nature: any gain by one was achieved at the expense of the other. When setting oil prices became the prerogative of OPEC alone, member nations took pricing decisions in an arena, or decisionmaking context, that excluded the providers of the benefits they sought. Within a distributive arena, benefits from increased prices are essentially "free." Higher prices to oil producers are not paid by other oil producers but by oil consumers. Thus, the arena also influences the content of policies—oil prices are higher since pricing policy has been made in a distributive arena. Policy content does determine which interests are affected, but this is no guarantee that each interest will have a seat around the bargaining table.[45]

In order to focus on both context and content in this analysis, the consumption dimensions developed by public finance theorists to type policies and arenas will be used.[46] The assumption underlying this choice is that policy is made by individuals who either themselves "consume" or benefit from their choices, or who represent directly the interests of policy consumers. The context and content of each policy interact according to who is represented in the decisionmaking process and how the policy outcome allocates costs and benefits across the policymakers or their constituents.

Figure 3 is a diagram of policies and arenas that uses the consumption dimensions to type them. The private goods cell of the typology corresponds to a distributive arena in Lowi's sense of the term.[47] This arena is one in which material benefits whose costs are either negligible, invisible, or borne by outsiders are allocated to individual actors directly or as representatives of the policy's target groups. In the OPEC context, increased income or control over oil resources is achieved by OPEC members at the expense of outsiders: multinational oil companies and oil consumers. A characteristic bargaining tactic in the distributive arena is *logrolling*, where actors band together in mutual support of a package of individually targeted benefits that are not mutually exclusive. OPEC has approximated logrolling behavior via the limited consensus with permissive implementation. This allows individual members to seek greater benefits at the expense of oil companies and oil consumers while anticipating support from OPEC, particularly in the event of

Figure 3: *A Typology of Policies According to Their Nature in Consumption, with the Arenas Associated with Policy Formation in Each Policy Category*

	Rival	Nonrival
Exclusive	Private Goods Distributive Arena	Exclusive Public Systems Pragmatic Arena
Nonexclusive	Common Goods Redistributive Arena	Pure Public Goods Public Interest Arena

their success. Although this is not logrolling in the strict sense of prior mutual support for specific individual benefits, in practice it amounts to much the same thing. The key to maintaining a distributive arena lies in seeking benefits whose costs are paid by outsiders. This minimizes conflict within the arena itself.

Cell II describes exclusive public systems. These systems are elaborate structures that provide highly desired benefits to policymakers and their constituents.[48] Public systems are interesting because other policy arenas and types are prominent in their formulation and implementation. Public systems are established to meet some "public interest" objective; for example, the OPEC information collection system was set up to provide a greater pool of information to its members than they might have been able to achieve singly, which then allows them to achieve greater policy coordination through tacit collusion. Implementation of public systems often involves redistribution among policymakers or their constituents. In our example, OPEC members willing to divulge important information about their production capacity or oil reserves subsidize the rest. Public systems policymaking is usually characterized by low levels of conflict because such policy objectives are highly desired. Even though costs may be disproportionately shared, that often reflects the relative usefulness of the policy to the contributors.[49] When conflict does arise in public systems provision, it usually occurs as disagreements with implementation schemes—what kind and at what cost—rather than over policy objectives.

Cell III describes common goods benefits. These are allocated in a redistributive arena where cost considerations are as much an object of bargaining as the amount and distribution of benefits. A redistributive arena is necessarily divisive because the relevant actors are (or represent) both benefactors and beneficiaries. In this context, bargaining is truly two-sided and zero-sum. Any benefit accruing to one actor must be paid for by the others. The division of production shares within OPEC is an example of a redistributive policy. Because it is inherently divisive, OPEC members have been reluctant to press hard for a production-sharing agreement. While the distributive bargaining mode is still able to generate sufficient benefits to justify the maintenance of OPEC, they do not have to. When these benefits appear to be threatened, prorationing becomes a high priority issue.

OPEC also generates a benefit of the pure public goods type by fixing the world market price for petroleum.[50] This kind of benefit, represented in Cell IV of Figure 3, is provided by actors bargaining in a public interest, or community of interests, arena. The concept of public interest is based on the kinds of notions of group solidarity and general welfare that underlie national defense policies in the domestic context. As in the national defense situation, the public whose interest is being furthered by this type of policy is limited to those represented in the policymaking arena. The distinguishing characteristic of a pure public good is that it is freely available to all actors or target groups once it is provided for any one of them, thus describing the behavior of non-OPEC oil exporters, such as Canada or Norway.

Within the framework of this typology we have examined the three types of bargaining or policymaking characteristic of OPEC and some of the policies associated with each type. Apart from its information-sharing system, OPEC does not produce public systems as a way of implementing its other policies. Internal bargaining in OPEC is thus confined to three arenas: the distributive arena, associated with individual income and power gains made at the expense of outsiders; the redistributive arena, associated with periodic attempts to arrive at a production-sharing agreement; and the public interest arena, where group solidarity motivates price setting and encourages all members to adhere to the pricing decision from which all members benefit.

The viability of OPEC is most severely tested when internal bargaining is redistributive. Serious redistributive bargaining, that is, bargaining that results in eventual resolution of substantive issues, has been minimal within OPEC, not only on the production prorationing issue but in the area of cartel leadership as well. This is evident in unresolved conflicts over pricing, such as the one that occurred at the Doha Conference of December 1976. It resulted in a formal two-tier pricing system that persisted for six months. If the Saudis did have enough market power to influence pricing unilaterally, they did not use it to force the rest of OPEC to charge the lower price that they favored.[51] United States officials feel that increased United States demand for oil in 1977 (due to the cold winter and the natural gas shortage) frustrated a Saudi bid for price leadership in the cartel.[52] However, it is also possible that the Saudis were not intending to force the rest of OPEC to follow them, but were instead warning other cartel members to be more sensitive to the position of the minority in OPEC favoring a moderate stand on oil price increases by making OPEC disunity public.[53] Whatever the motivation, there is no evidence that the Saudis attempted to force the price structure lower than the second tier level. Such an action would have been a major assault on the price structure, something the formal two-tier system was not.

OPEC has been able to avoid conflict over cartel benefits when member national income levels remained high enough to avoid a decision on market shares. Income maintenance has been accomplished through pricing policy. Price increases to expand the size of the OPEC pie are possible as long as demand does not decline below a level capable of sustaining minimum desired income levels to producers with excess production capacity, and as long as oil prices remain competitive with the prices of acceptable substitutes. Market share allocation becomes a bargaining priority in OPEC when demand for OPEC oil declines.

Maintaining a Distributive Arena

The dominance of the distributive mode in OPEC bargaining is reinforced by the existence of multiple, noncoincident policy goals among its members. This enables OPEC to assemble policy "packages" that include a variety of outcomes valued by members. Since the material benefits of income from petroleum sales are limited by total world demand for oil, having nonmaterial benefits such as production control increases the likelihood of maintaining a distributive bargaining arena by providing additional benefits to distribute. Particularly where material benefits are insufficient to meet some basic, intra-OPEC level of total demand for national income, alternatives to income as a desired policy outcome are imperative. The existence of political goals valued by Arab members of OPEC permitted the assembling of a production cutback/major price increase policy package in late 1973. Individual national conservation policies that coincided with the inability or unwillingness of other member nations to reduce their oil incomes during the first post-embargo recession supported the price structure during a period of falling demand.

The mere existence of multiple petroleum policy goals will not preserve a distributive arena for OPEC bargaining if these goals are not also mutually compatible. As a result of their increased incomes, OPEC members are becoming increasingly differentiated economically as each follows its own path to development. This differentiation contributes to incompatibility within OPEC where the basis for solidarity has historically been economic rather than political. For example, extensive investments in consumer nations have made some OPEC members oppose others on price increases that could weaken consumer country economies. In cases where multiple policy outcomes are mutually exclusive, bargaining shifts to a redistributive mode reflecting the zero-sum nature of value allocation. In this way the diversity in OPEC, which is a source of its strength, could threaten continued success by transforming bargaining on issues that had been treated in a distributive arena into redistributive conflicts among OPEC members.

OAPEC in OPEC

The most powerful role for OAPEC within OPEC would be control of the cartel. In effect, OAPEC would be the real petroleum cartel. If this were so, OAPEC could set prices and production levels to suit itself, regardless of what the non-Arab members of OPEC might wish to do. In order to see if OAPEC controls OPEC, it is necessary to assess its potential for cartel leadership and to determine what OAPEC, or its members acting en bloc, have actually done to demonstrate their leadership or domination of OPEC. Potential to control OPEC can be seen by measuring the degree of OAPEC's control of OPEC petroleum resources. Actual control would be indicated if OAPEC members form a coalition that successfully pursues collective Arab goals through OPEC policy. OAPEC solidarity could be a source of divisive conflict within OPEC or it could function to aggregate a subset of members with differing though compatible policy aims from the rest of OPEC, thus contributing to OPEC's solidarity by helping to preserve its distributive bargaining arena. Finally, if OPEC and OAPEC are regarded as interacting entities, we need to know if they are interdependent, or if one is dependent on and thus subordinate to the other.

The Control of OPEC

It is tempting to regard OAPEC members as dominant in OPEC because of the magnitude of the oil resources they control. Figure 4 shows estimates of crude oil reserves for principal OPEC members from 1973 to 1979. Figure 5 shows crude oil production rates from 1973 to 1979. Official oil reserves are a form of propaganda put out by oil companies and oil-exporting countries in order to influence both outsiders and insiders to OPEC politics.[54] However, most reserves estimates are probably correct in showing the relative importance of oil-exporting country holdings. If the reserves of the OAPEC/OPEC members are compared to reserves in the rest of OPEC, the results are as shown in Table 17. During this recent period, OAPEC members controlled roughly three-fourths of OPEC's petroleum reserves, potentially decisive in exercising control within OPEC over a long period as production draws down reserves.

Production is an important factor in cartel control; it is the measure of market dominance and therefore realized control. Table 18 shows the collective production of OAPEC-OPEC members over the same period described in Table 17. From this we can see that OAPEC is responsible for about

Figure 4: *Estimated Crude Oil Reserves in Selected OPEC Countries, 1973-1979 (year-end totals)*

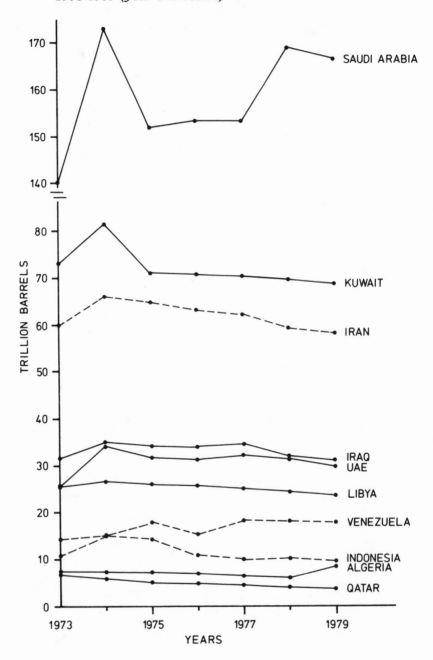

**Figure 5: *Crude Oil Production in Selected OPEC Countries, 1973-
1979 (year-end totals)***

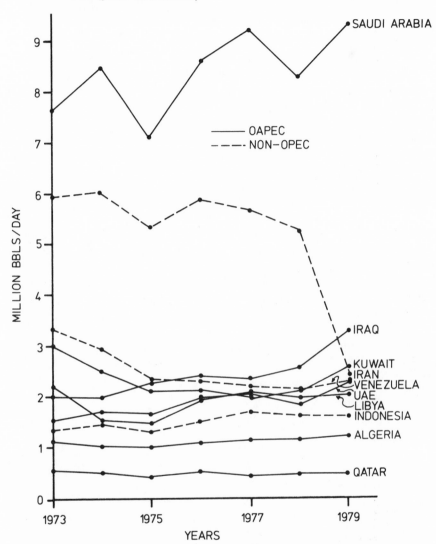

two-thirds of current OPEC production, again probably enough to control
pricing policy, particularly with regard to cutting back production to affect or
maintain price increases. For example, during the 1980 oil glut, OAPEC
members cut back production in the spring to narrow the gap between supply
and demand. Kuwait cut its production 25 percent to 1.5 million barrels per
day (MBD) and Libya cut its production 17 percent, from 2.1 MBD to 1.75

TABLE 17 OAPEC Oil Reserves

	OAPEC Reserves[a,b,c]	As % of OPEC Reserves
1973	355,068	76.3
1974	310,140	73.5
1975	363,820	75.0
1976	328,320	73.1
1977	326,900	74.5
1978	327,325	74.4
1979	336,396	75.6

Source: OPEC, Annual Report, 1978, p. v.

Note: Reserves expressed in millions of barrels. OAPEC reserves calculated as the sum of reserves of Arab OPEC members.

[a]Excluding Egypt, Syria, and Bahrain.
[b]As of January 1 of each year.
[c]Low estimates for Iraq.

MBD.[55] Algeria, a country that has a high capital absorption capacity, also cut production by 20,000 to 30,000 barrels per day (B/D), with the possibility of withdrawing an additional 1,000,000 B/D from the export market to use as feedstock in its own refineries.[56] Because Saudi Arabia produced 1,000,000 B/D over its limit throughout the first two quarters of 1980, other OPEC members must cut back even more in order to keep prices up, and the decision by Algeria was significant in this regard. However, even for Algeria, production cuts were not so damaging financially in 1980 as they would have been two years before because of the doubling of oil prices during those years.

TABLE 18 OAPEC Oil Production

	Total OPEC	OAPEC	OAPEC/OPEC[a]
1973	11.3	6.5	57.5
1974	11.2	5.5	49.1
1975	9.9	5.6	56.6
1976	11.2	6.7	59.8
1977	11.5	7.0	60.9
1978	10.9	6.7	61.5
1979	11.2	7.6	67.9

Source: Petroleum Economist, June 1980, p. 273.

Note: Production expressed in trillions of barrels.
[a]Calculated.

Reserves levels indicate long-run market power while production levels show short-run or realized market power. These two measures are interrelated, for current market power is exercised at the expense of future production. The reserves to production ratio (R/P) is a guide to the future status of producers. It reflects the individual producer's potential to maintain its current status or to raise it over time. Table 19 shows R/P values for most OPEC members for six recent years. The table reflects the instability of reported oil reserves and fluctuations in yearly oil production. Each figure can be read as the approximate number of years more the country can produce the same amount of oil as it did in that year. The effect of petroleum conservation policies on the conservation of power in OPEC is evident from the general trends for Venezuela and Kuwait. In 1973, Venezuela was producing at a rate that would exhaust its reserves in about twelve years, if it were maintained. The government of Venezuela cut back production as part of a resource conservation policy. The result has been to extend Venezuela's oil-producing future by over ten years. The effect of conservation on an oil-producing country's production life span is even more vividly evident for Kuwait, which has extended its producing future to over ninety years[57] (see Figures 4 and 5).

TABLE 19 *R/P Ratios for Selected Members of OPEC*

	1973	1974	1975	1976	1977	1978
Saudi Arabia[a]	50.7	55.9	59.8	48.8	45.6	55.9
Kuwait[a]	65.8	87.6	93.6	89.6	97.4	90.7
Iraq[b]	43.9	51.4	41.8	38.9	37.9	33.4
Qatar	31.3	31.7	36.6	32.0	35.2	22.7
Libya	32.0	48.3	47.4	36.4	33.0	33.7
UAE	45.9[c]	55.3	51.2	43.8	44.2	46.9
Algeria	19.5	20.6	21.4	17.2	16.5	14.1
Iran	27.9	29.9	32.8	29.1	29.8	30.9
Venezuela	11.4	13.8	20.7	18.3	22.3	22.8
Indonesia	21.5	29.9	29.1	19.2	16.2	17.1
Nigeria	26.6	25.4	31.0	25.7	24.4	26.1

Sources: Reserves data from OPEC, *Annual Report, 1978*, p. v, year-end figures; production data from *Petroleum Economist*, June 1980, p. 273.

[a]Includes one-half Neutral Zone reserves.
[b]Low estimates of reserves.
[c]Data for Abu Dhabi and Dubai only.

If producing countries are to control oil prices they must be able to manipulate supply. Those OPEC members that can measurably affect global oil supply are those with the highest production capacity. If they choose to hold back production they can push prices upward. Supply cuts are more easily tolerated by countries with budget surpluses, such as Saudi Arabia or Kuwait, or by those that have a relatively low degree of dependence on oil sales to maintain a national standard of living, such as Bahrain. Downward influence on prices is obtained by increasing the total supply of oil in the market. Excess or reserve production capacity is the critical factor needed to increase supply in the short term. Tables 20 and 21 show the distributions of production capacity and excess production capacity respectively for all OPEC members. Looking first at Table 20, the preeminent role of Saudi Arabia in production is obvious. If its capacity to produce over a fourth of OPEC's total crude oil is added to the production capacity of the other Arab OPEC members, the result is that OAPEC members control about two-thirds of OPEC's production capacity.

TABLE 20 OPEC Oil Production Capacity, 1979

	Capacity in MBD	
Country	Fesharaki[1]	Energy Index[2]
Saudi Arabia[a,b]	10.80	11.5
Iran	4.50	6.7
Iraq	4.00	4.0
Kuwait[a]	2.80	3.0
UAE	2.48	} 3.0
Qatar	0.65	
Venezuela	2.40	2.8
Nigeria	2.40	2.6
Libya	2.10	2.5
Algeria	1.20	1.8
Indonesia	1.65	1.7
Ecuador	0.25	} 0.4
Gabon	0.25	
TOTAL	35.48	40.0
Domestic consumption	2.30	2.3
Available for export if production maintained at capacity	33.18	37.7

Sources:
[1]Fereidun Fesharaki, "Global Petroleum Supplies in the 1980's: Prospects and Problems," *OPEC Review* 4 (Summer 1980), 33.

[2]Environment Information Center, *The Energy Index 79* (New York, 1979), 78.

Notes:
[a]Includes one-half Neutral Zone Capacity.

[b]Walter J. Levy agrees with the lower figure for Saudi Arabia (telephone interview, 28 July 1980).

Prior to the Iranian revolution, OAPEC members also controlled about three-fifths of OPEC's *excess* production capacity. In 1979, the proportion of excess capacity controlled by Arab oil producers had gone to about half of all OPEC excess capacity. The position of Saudi Arabia is particularly relevant, for, as we shall see, it was the only country attempting to hold prices down in 1979 during what was at first perceived, incorrectly, as a supply shortfall.

TABLE 21 *Excess Oil Production Capacity, 1979*

Country	Estimated Excess Capacity in MBD	
	Fesharaki[1]	*Energy Index*[2]
Saudi Arabia[a]	1.32	2.02
Iran	1.52	3.72
Iraq	.62	0.62
Kuwait[a]	.55	0.75
UAE	.66	}0.68
Qatar	.15	
Venezuela	.06	0.46
Nigeria	.07	0.27
Libya	.04	0.44
Algeria	.06	0.66
Indonesia	.05	0.10
Ecuador	.04	0.0
Gabon	.04	0.0
TOTAL	5.18	9.72
OAPEC/OPEC	65.6%	52.2%

Sources: Production estimates from Fesharaki.

[1]Fereidun Fesharaki, "Global Petroleum Supplies in the 1980's: Prospects and Problems," *OPEC Review* 4 (Summer 1980), 33.

[2]Environment Information Center, *The Energy Index 79* (New York, 1979), 78.

Note:

[a]Includes one-half Neutral Zone Capacity

The potential to control oil prices may not actually have rested with the Arab oil producers even before 1979, as historical and statistical analyses of events seem to show.[58] In a series of factor analyses of nineteen indicators of energy and political characteristics of OPEC members, Charles F. Doran obtained factors representing possible coalitions of OPEC members based on their similarity on the chosen indicators.[59] He found no OAPEC coalition in any of the years in his series (1969-75), although a factor containing all four conservative-regime OPEC/OAPEC members was the principal coalition every year except 1973, the year of the oil crisis.[60] An Arab coalition without revolutionary-regime members cannot represent Arab policy nor can it represent OAPEC in view of the even split in its membership along the lines of regime characteristics.

Table 22 shows regime dissimilarity in OAPEC and some of the other differences that make it hard for OAPEC members to achieve unity on oil-pricing policy. A traditional-regime group composed of Saudi Arabia, Kuwait, and the United Arab Emirates is characterized by moderate to high levels of oil reserves, a low immediate need for capital and basically one-industry economies. These countries tend to prefer relatively lower prices for oil, both to discourage substitutes for OPEC oil that could affect their future income and to protect their investments in consumer nations whose economies would be adversely affected by higher energy prices. Despite its low-level reserves, Qatar, with its tiny population, usually shares the price preferences of the other conservative members of OAPEC. These four nations were the ones Doran found in his dominant coalition within OPEC.

At the other extreme, Algeria, Egypt, Syria, and Iraq have great immediate need for capital for development and reportedly low to moderate oil reserves. This disposes them to favor relatively higher oil prices to get as much immediate income as possible before sustitutes for OPEC oil can depress prices by reducing consumer demand.

Two OAPEC members share attributes of both groups. Bahrain is the most developed member of OAPEC. It has produced oil since 1932, and has used its income over the years to diversify its economy and decrease its dependence on oil for current income.[61] Libya's oil policy has always been unique among the Arab states.[62] The regime has a political affinity with the other revolutionary-regime members of OAPEC, but its low need for capital, modest pre-oil economic diversity, and moderate reserves level decrease its economic (although not its political) interest in high oil prices. Within a range bounded on one hand by the moderates' ideal price and on the other by the ideal price of the radicals, Bahrain and Libya are economically indifferent to oil prices relative to the other two groups.[63]

Other analysts have also developed categories for typing OPEC members on the price issue. Øystein Noreng looks at OPEC as being made up of three factions, one that contains countries that produce oil near or at their production capacity for income needs; a second that includes countries that produce oil according to their views on conservation and their preferences on prices; and Saudi Arabia, price leader and ultimate residual producer.[64] Noreng sees behavior in OPEC as complex, rather than as a simple function of oil reserves and income needs. For example, if Saudi Arabia wants prices to hold or to fall, Noreng says it can moderate prices by increasing production, thus forcing the countries that produce for income to maintain capacity production to avoid losing income. From the other side, if Saudi Arabia wishes to raise prices, it can do so, reduce its own production, and enable the producers-for-income to lower their production too because at the higher prices they can meet their income requirements at less volume.[65] As we have seen, however,

TABLE 22 Attributes of OAPEC Members

	Regime Type[1]	Oil Reserves[2]	Need for Capital[3]	Income Dependence*
Algeria	Socialist republic	low	high	moderate to high
Bahrain	Islamic sheikhdom	low	moderate	low to moderate
Egypt	Socialist republic	low	high	low to moderate
Iraq	Socialist republic	moderate	high	high
Kuwait	Islamic sheikhdom	high	low	high
Libya	Socialist republic	moderate	low	moderate to high
Qatar	Islamic sheikhdom	low	low	high
Saudi Arabia	Islamic monarchy	high	low	high
Syria	Socialist republic	low	high	low
United Arab Emirates	Federation of Independent Islamic Sheikhdoms	moderate	low	high

Sources:

[1] David Crawford, *Getting Work in OPEC* (London, 1975), various pages.

[2] American Petroleum Institute, *Data Book* (1977), Section XIV, Table 1. High=greater than 6^{10} barrels (bbls.); Moderate = 2^{10} to 6^{10} bbls.; Low = less than 2^{10} bbls.

[3] Capital absorption capacity: all data except that on non-OPEC members are from Doran, "Conflict, Cohesion." All else is from Crawford.

* Degree of dependence for national income on oil and gas sales as of 1979.

Saudi Arabia loses much of its ability to influence prices during real or perceived acute supply shortfalls. Noreng views conflict in OPEC as a function of world demand, excess capacity and its distribution across OPEC members, and the members' respective income requirements.

Paul Leo Eckbo has categorized OPEC members according to characteristics similar to those chosen by Doran and Noreng. His results give him three groups of oil producers. The "hard core" group includes countries with financial surpluses and excess production capacity. Eckbo names Saudi Arabia, Kuwait, the UAE, Libya, and Qatar as hard-core producers. The "price pushers" produce at or near capacity and can use all their income for current expenditures. The price pushers are Iran, Venezuela, and Algeria. Finally, Eckbo's "expansionist fringe" includes Indonesia, Nigeria, and possibly Iraq, countries with some excess capacity but also having the ability to use efficiently additional current income.[66]

None of the means for categorizing OPEC members has come up with a coalition that includes all OAPEC members. Even if Saudi Arabia is considered as a case by itself, the other Arab members of OPEC appear to be split in their preferences for various prices for oil. Even when the OPEC Conference can keep oil prices within a narrow range, Arab countries oppose other Arab countries on prices. Table 23 lists a series of intra-OPEC confrontations over oil prices running from March 1973, just prior to the Arab oil embargo, and extending through December 1977. A glance at the table shows consistent opposition between Saudi Arabia and Iraq, a significant split

TABLE 23 *OPEC Confrontations, 1973-1977*

Date	Bloc I	Bloc II	Issue
03/73[1]	Saudi Arabia	Algeria	Raise oil prices 10% (II)
05/73[2]	Saudi Arabia Abu Dhabi Iran Nigeria	Libya Iraq Kuwait	Support 12/72 participation agreements (I) or demand more from the oil companies (II)
12/73[3]	Saudi Arabia	Iran	Raise marker crude price to $11.651/bbl. (II)
06/74[2]	Saudi Arabia	Kuwait Iran	Freeze prices (I) Did increase royalties (II)
09/74[2]	Saudi Arabia Algeria* Kuwait* Iran*	Iraq Libya	Opposed to higher taxes or higher royalties (I)
03/75[2]	Saudi Arabia	Algeria	Call a special OPEC Conference to develop a position vis-à-vis Washington Energy Conference (II)

Table 23 continued

Date	Bloc I	Bloc II	Issue
09/75[2]	Saudi Arabia Indonesia* Libya*	Iran	Raise oil prices at least 15% (II)
11/75[2]	Kuwait	Iraq	Cut own oil prices (I)
05/76[4]	Saudi Arabia Nigeria Algeria* Iran*	Venezuela Iraq Libya	Freeze oil prices (I)
12/76[2]	Saudi Arabia United Arab Emirates Venezuela*	Iraq Libya Iran Kuwait	Raise prices 5% (I) Raise prices 10% (II)
06/77[1]	All but Bloc II	Iraq Libya	Raise prices 5% (II)
09/77[5]	Saudi Arabia Kuwait Algeria United Arab Emirates	Iraq Libya	Work for "price restraint" within OPEC (I)
12/77[6]	Saudi Arabia Iran Kuwait United Arab Emirates	Venezuela Iraq Libya	Price freeze (I)

Sources:
[1]*New York Times*, 16 March 1973, p. 1; 3 July 1977, p. E1.
[2]*Oil and Gas Journal*, various issues.
[3]Raymond Vernon, ed., *The Oil Crisis* (New York, 1976), pp. 68-69.
[4]*The Middle East* 21 (July 1976).
[5]*Houston Post*, 16 September 1977, p. E2.
[6]CBS TV "Evening News," 20 December 1977.

*Took an intermediate position on the issue listed.

given their common location in the Gulf and the fairly widespread belief that Iraq may eventually rival Saudi Arabia in the measure of its oil reserves.[67] Libya was frequently on the side of Iraq, urging higher oil prices. For the most part, all the African producers, Arab and non-Arab, have been militant in their demand for higher prices. Iraq was the anomaly among the Gulf producers, as much because of its political opposition to most of the other Gulf states as because of its reported reserves position and its income requirements. Iraq's more recent support of the Saudi position within OPEC has not resulted in any change regarding the absence of an OAPEC coalition.

Relative consistency in member positions in OPEC was reduced by the Iranian revolution, for political and economic reasons. In a 1977 estimate, the United States Central Intelligence Agency (CIA) not surprisingly predicted that when, eventually, the global supply of oil should fall below the level of demand, oil producers would be able to charge whatever they liked for their petroleum exports.[68] The CIA predicted that this would not happen until 1983, but falling Iranian production beginning in November 1978 created an oil supply panic similar to the one in 1973.[69] Total OPEC oil exports in 1978 were down 1.5 MBD from 29.4 MBD in 1977, but they increased by 0.8 MBD the following year owing to increases of 10 percent or more in total exports from Saudi Arabia, Iraq, Nigeria, Kuwait, and Ecuador.[70]

The Iran supply crisis came at the end of a year that had seen African OPEC members drop the prices of their crudes because of competition from North Sea oil, which reduced their oil exports.[71] In spite of the competition, OPEC members did decide to increase oil prices a modest 10 percent at the end of 1978 because of declining real prices for their oil and relative increases in import prices. They instituted a new method for applying increases, by splitting the total increase into four parts to be phased in quarterly. For the first time since October 1973, Iran sided with Saudi Arabia and the other OPEC moderates on the size of the price increase.[72]

Given the small size of the price increase, coming on top of high inflation in OECD countries and a corresponding decline in the dollar, the official currency for making payments to OPEC, OPEC members were shocked at the outburst of condemnation from oil importers when the price increase was announced. Government officials in the United States, Europe, and Japan predicted that the OPEC action would increase worldwide inflation and impede global economic recovery. United States Energy Secretary James Schlesinger said he thought that the price increase would even interfere with security arrangements between producer and consumer governments.

OPEC price radicals and moderates were equally disturbed at the strident consumer response to the 10 percent increase. Oil and finance ministers from most OPEC countries spoke publicly in defense of their action. UAE oil minister, Mana Saeed al-Otaiba, said, "We were astonished by the reactions of the industrialized countries, particularly because the oil price increase decision was reasonable and should not inflict any damage on the world economy."[73] The new OPEC Secretary General, Rene G. Ortiz of Ecuador, warned consumer governments that OPEC might decide on even further increases in June 1979 if inflation and the dollar decline were not halted.[74] The oil-exporting governments and OPEC itself were united in their perception that the price increase they had decided upon in December 1978 was justified. They were also united by their anger at what was widely perceived as consumer overreaction to it.

Less than a month from the time that the price increase was announced, the Shah was ousted from Iran. Iranian oil production continued its precipitate decline creating an immediate short-term supply shortage of about 2 to 5 percent. Oil prices in the spot market took off in January. Prices for the average crude increased by about $4.00 per barrel by the end of the month (see Table 24). The price explosion in the spot market was followed by unilateral increases in official prices. The Saudi increase was the smallest. Their production ceiling was raised 1 MBD during the first quarter of 1979 in response to the crisis, and Aramco was charged the scheduled 1979 fourth-quarter prices, about $1.00 more per barrel, but only for the extra oil. Abu Dhabi and Qatar followed with $1.00 per barrel increases on their principal crudes beginning 15 February, Libya raised its official selling price by $0.68 per barrel, and Kuwait and Venezuela announced an increase of $1.20 per barrel at the end of the month.[75] During the first quarter of 1979 Iraq refused to raise official prices for its oil, asking instead for "voluntary" premiums from its customers.[76] By the end of the first quarter of 1979, OPEC price coordination had disintegrated.

TABLE 24 *Spot Market Premiums for Selected OPEC Crudes, January 1979*

| | Official Price | Premium | | |
		Early January	Mid-January	End January
Arabian Light*	$13.34	$1.41	$2.26	$3.16
Kuwait	12.83	.67	1.67	3.67
Basrah Light	13.29	1.46	2.41	4.21
Murban	14.10	n.a.	n.a.	3.90
Zarzaitine	14.75	2.25	3.25	5.25
Brass River	14.84	2.41	3.41	5.16
Es Sider	14.52	1.48	3.48	3.98

Source: Middle East Economic Survey, 5 February 1979, p. 2. Where a price range was indicated, the lower price is reported here.

*Pre-June 1980 Marker Crude.

   An Extraordinary Meeting of the OPEC Conference was called in Geneva at the end of March to deal with the price situation. Prior to the Conference meeting, Abdul Amir al-Anbari, chair of the Iraq Fund for External Development, published an article containing a modest proposal for future OPEC oil price policies.[77] Anbari suggested that OPEC set a price range rather than a single price for the marker crude and allow individual governments to make adjustments in their oil selling prices within the official limits in response to market changes occurring between OPEC meetings. Market pressures were too strong to permit even this innovative proposal to be adopted. In essence, OPEC members agreed to disagree on everything but a new price for the marker crude, which was to rise to the planned fourth-quarter level immediately. The surcharges which individual members had been adding to their official prices remained uncontrolled.

Variations in the actual selling prices for OPEC oil widened after the Conference meeting. Saudi Arabia continued to charge only the marker price for all oil except its Berri crude, which carried a quality premium. Other Gulf producers raised their surcharges, and the African producers (Nigeria, Libya, and Algeria) raised their quality premium $2.50 per barrel to $4.00, making their oil the most expensive in OPEC. Kuwait raised its surcharges and its base price and formally instituted contract premiums of about 9 percent for supplementary crude.[78]

The agreement to disagree fit into OPEC's traditional pattern of allowing individual members some latitude to improve individual postions vis à vis a basic consensus decision. Even so, the nonpolicy that resulted did little to resolve continuing member conflict over prices. Saudi Arabia maintained a low increase position even after the Egyptian-Israeli peace treaty was signed and its relations with the United States hit another new low. Iraq and the smaller Gulf states took an intermediate position. Iraq had been the old opponent of Saudi Arabia in OPEC price meetings since 1974, and a political opponent in OAPEC and other international organizations for much longer. The Iranian revolution and the Arab decision to isolate Egypt after the treaty signing brought Iraq and Saudi Arabia closer together. This rapprochement coincided with a distinct moderation of Iraq's attitude toward oil prices beginning in early 1979. The African states, as usual, maintained their high price positions. The quality of African crudes has traditionally supported a differential in prices between them and the Gulf crudes, and the governments of Libya, Nigeria, and Algeria anticipated that their much higher prices would hold.

As spot market prices continued to rise, base prices, differentials, and surcharges increased at a dizzying pace. By the end of May, Algeria had increased its official price to $21.00 per barrel, then the highest price in OPEC. Alongside price escalation in OPEC, non-OPEC members followed their usual pattern of matching or exceeding OPEC prices. In June, other OPEC members outstripped Algeria. Iraq began including a "most favored seller" clause in new contracts beginning 1 June 1979. The clause required crude oil customers to pay Iraq "the highest premium charged by any other OPEC producer" over and above the base price for any Iraqi crude.[79] Kuwait soon followed in establishing the most-favored-seller principal in its contracts.

Officials in the Gulf oil countries regarded most-favored-seller clauses as leading to price reunification. African producers naturally disagreed because of quality differences between Gulf and African crudes. Differences over differentials were the main conflicts at the June 1979 OPEC Conference meeting, and, as in March, no agreement could be reached. However, the Conference did set a ceiling of $23.50 per barrel on crude oil prices. Individual OPEC countries then set their own prices relative to the ceiling and ascribed the resulting price totals to creatively combined packages of surcharges and differentials,[80] setting the stage for a new round of increases later.

Although agreement on a ceiling price was made on the assumption that no member country would raise production levels in a material way, Saudi Arabia did raise production levels by 1 MBD in July, as it had in the first quarter of the year, to compensate for production shortfalls in Iran. The government justified its move as necessary to fund projects "vital . . . to the well-being and prosperity" of its population under the current five-year plan.[81] At the time, spot market prices were about $10.00 per barrel over the official ceiling, which probably also affected the Saudi decision. Spot market pressure on non-OPEC contract prices was evident later in July when Egypt signed a contract for its high quality Suez Blend at a price of $32.50 per barrel, close to the spot price.

In spite of the actions of Saudi Arabia and Egypt, the oil market remained fairly calm until Kuwait raised the price of its Kuwait crude in October. As far as the Kuwait government was concerned, its decision was intended solely as a readjustment to restore the traditional differential between Kuwait and other Gulf crudes.[82] However, the government also reinstituted the most-favored-seller clause, temporarily obsolete under the June price agreement, erasing some differentials. Leapfrogging began almost immediately, Libya openly bursting through the $23.50 per barrel ceiling by setting a $26.27 per barrel price on its Zuetina crude. Spot prices also shot up, and there were reports that Japanese buyers were taking crude directly from OPEC governments at prices in excess of $40.00 per barrel, even higher than spot prices.[83]

In December, shortly before the OPEC Conference was scheduled to meet in Caracas, Saudi Arabia, Qatar, the UAE, and Venezuela announced a $6.00-per-barrel price increase. The move was widely reported as a preemptive bid to reunify OPEC prices. It also enabled Saudi Arabia to punish the Aramco partners for taking huge profits from selling relatively cheap Saudi crude at world prices, by making the new price effective from 1 November. The day the increase was announced, 15 December, was the last date on which retroactivity could have been imposed under the payment system then in effect.[84]

It was reported that Kuwait and Iraq were initially part of the Saudi-led move to reunify prices but backed out because of disagreements with Qatar and the UAE over differentials.[85] Clashes over differentials also took place in the Caracas meeting of all the OPEC members, who could not agree on either differentials or a marker crude price. Even so, some OPEC members thought that after Caracas prices would approximate another two-tier system, with the Africans and Iran selling at the top price and the other nine selling for the second-tier price.

By the end of January 1980, the new year began to look very much like 1979. An attempt was made by Saudi Arabia to realign itself with the moderate coalition by raising the marker price by $2.00 per barrel. The attempt was foiled by Kuwait's countermove to raise its own price by $2.00 per barrel. Kuwait's lead was followed by similar price increases in Iraq, Qatar, and the UAE. This left Saudi Arabia in the same relative position in early February as it had been in a month before.

Back in January 1979, before oil prices began their alarming spiral upward, Saudi Arabian Oil Minister Yamani pointed out that his government could not moderate prices in a tight supply situation.[86] Saudi production capacity, earlier predicted to reach 14.5 MBD by 1980, was actually closer to 11.5 MBD in 1979, not enough to push prices down. Production shortfalls upset the world oil market as much for pyschological reasons as because of real shortages in consumer markets. It has become increasingly likely that small shortages, quickly compensated for by production in other countries, will continue to have as much power to create panic in the market as an absolute shortage of oil.[87]

A look at oil prices through the period just discussed shows the price coalitions in OPEC. December 1978 and January 1979 prices were set before the Iranian oil shortage was reflected in the spot market. The January prices included the first of the quarterly increases decided upon by the December 1978 OPEC meeting. The July prices were set after the disorderly market of the first half of 1979 shot prices upward, in conformity with OPEC's June decision to set a $23.50 ceiling on prices. The December prices showed a return to individual price setting after the Kuwait increase in October and subsequent leapfrogging. Some prices increased again in January 1980, reflecting OPEC's inability to unify prices the month before.

Differences in oil prices due to quality and location are reasonable and customary, and in Table 25 can be readily seen within countries on prices for more than one crude, and occasionally between countries, such as Abu Dhabi and Qatar. Refusal to conform to the traditional differentials creates price leadership behavior between sellers of similar crudes and between the African and the Gulf producers. During the times when an OPEC price decision, either on the marker crude or on a ceiling, was adhered to, prices varied moderately, generally following traditional differential patterns. The African producers tend to be the price leaders during disorderly markets, pulling the Gulf producers behind them as Gulf prices hurry to close the gap between customary and actual differentials and African prices zip ahead whenever the gap narrows.

Since the Iranian revolution, Iran has been part of the African group working to push offical OPEC-set prices higher, but not always able to get the highest prices for its own oil because of its lower quality and less convenient location with respect to the European Market. Because of the price increases of 1979-80, however, the Iranian government was able to make the same amount of money from its oil in 1979 as it did in 1978, even though it sold only half as much. From the point of view of oil consumers, this illustrates the

TABLE 25 *Official OPEC Crude Oil Sales Prices ($/bbl.) End December 1978,*
Beginning January 1980

| | Sales Prices | | | | % Increase 12/78 |
	12/78	1/79	7/79	12/79	12/79
Abu Dhabi					
Zakum 40°	$13.17	$14.01	$21.46	$27.46	108%
Murban 39°	13.26	14.10	21.56	27.56	108
Umm Shaif 37°	13.04	13.78	21.34	27.36	110
Saudi Arabia					
Light 34°	12.70	13.34	18.00	24.00	89
Medium 31°	12.32	12.89	17.54	23.54	91
Heavy 27°	12.02	12.51	17.17	23.17	93
Iran					
Light 34°	12.81	13.45	22.00	28.50	122
Med/Hvy 31°	12.49	13.06	19.90	27.77	124
Iraq					
Kirkuk 36°	12.88	13.52	22.00	21.18	72
Basrah 35°	12.66	13.29	19.96	21.96	73
Kuwait					
Kuwait 31°	12.22	12.83	19.49	21.43	75
Qatar					
Dukhan 40°	13.09	14.03	21.42	27.42	109
Marine 36°	13.00	13.77	21.23	27.23	109
Algeria					
Saharan					
Blend 44°	14.10	14.80	23.50	30.00	113
Zarzaitine 42°	14.05	14.75	23.45	30.00	114
Libya					
Brega 40°	13.85	14.69	23.45	29.95	118
Sarir 38.5°	13.29	14.13	22.90	29.40	123
Nigeria					
Light 37°	14.12	14.82	23.49	29.97	111
Indonesia					
Minas 35°	13.55	13.90	21.12	23.50	73
Venezuela					
Officina 34°	13.99	14.69	22.45	26.75	91
Tia Juana 24°	12.39	13.01	18.68	22.93	85

Source: Petroleum Economist (January 1980), p. 3.

vicious cycle of higher prices leading to lower production which in turn supports higher prices, and so on (see Table 26). This cycle did not begin with the 1973-74 price rises, in spite of dire predictions by Western observers that it might. But 1979 marked a new era in the world oil market, and since January of that year OPEC crude oil prices have appeared to be largely divorced from supply and demand constraints.

The June 1980 meeting of the OPEC Conference in Algiers has been called "a step toward price reunification" by an OPEC official[88] and a "tentative step" by the MEES news editor.[89] These judgments may be premature, but one aspect of the Conference's price decision appears to have weakened the ability of Saudi Arabia to influence prices. From 1973 to 1979, OPEC's pricing decisions amounted to setting a price for 34° API Arabian Light oil from Saudi Arabia. In June 1980 OPEC officially abandoned Arabian Light as its marker crude and established "a marker crude which has the same characteristics at those of Arabian Light without necessarily reflecting its price."[90] According to an OPEC official, the decision to create a "theoretical" or "hypothetical" marker crude was taken because the official price of Arabian Light "came to be decided by the government of Saudi Arabia...no longer by a collective decision of the OPEC Conference."[91]

While the government of Saudi Arabia continues to keep the price of Arabian Light below the other marker price, it is setting a floor price for OPEC oil rather than influencing price ceilings. Political pressure within OPEC is thus directed more strongly toward Saudi Arabia to increase the price of Arabian Light, rather than by Saudi Arabia toward the African producers to keep their crude prices down. A below-marker price for Arabian Light after the June 1980 decision led to greater than usual differentials for the African producers based on the selling prices of Saudi crudes. The subsequent decision by the OPEC Conference, meeting in Bali in December 1980, to establish a price band for the marker as well as for other crudes, has not altered the situation. Instead, it has made OPEC pricing even more chaotic than before by allowing each producer to set the prices for its crudes based on a marker price of its own choosing.

As OPEC relinquishes much of its control of oil prices to its members, the ability of any coalition to exert a deciding influence on oil prices decreases. The Arab members of OPEC never controlled OPEC prices. The joint membership of Arab oil ministers in the OPEC Conference and the OAPEC Council led to policy coordination in 1973 between the OPEC price increase of 16 October and the Arab oil embargo agreed upon the following day. By December 1973, however, Saudi Arabia had broken with the Arab and non-Arab OPEC members pushing for even higher oil prices, charting a frequently lonely course aiming toward the stabilization of oil prices close to the December 1973 nominal rates.

TABLE 26 Oil Exports[a] and Revenues[b]

	1974		1977		1978		1979		1978/1979	
	Exp.	Rev.	Exp.	Rev.	Exp.	Rev.	Exp.	Rev.	Exp.	Rev.
Saudi Arabia	8.5	22.6	9.0	38.6	8.1	34.6	9.3	57.7	+15%	+67%
Iraq	1.8	5.7	2.2	9.8	2.4	9.6	3.3	23.4	+37	+144
Iran	5.7	17.5	5.0	21.6	4.5	20.9	2.4	20.8	−47	−2
Libya	1.5	6.0	2.0	8.9	2.0	8.6	2.0	16.3	0	+90
Nigeria	2.2	8.9	2.0	9.6	1.8	8.2	2.1	16.1	+17	+96
Kuwait	2.4	7.0	1.9	7.9	2.1	8.0	2.4	16.0	+14	+100
UAE	1.7	5.5	2.0	9.0	1.8	8.0	1.8	12.8	0	+60
Venezuela	2.8	8.7	2.0	6.1	1.0	5.6	2.0	12.0	+5	+114
Algeria	0.9	3.7	1.1	4.3	1.1	4.6	1.1	8.8	0	+91
Indonesia	1.2	3.3	1.5	4.7	1.4	4.8	1.3	8.1	−7	+69

Source: Oil and Gas Journal, 28 April 1980, p.47. Data originally from Bankers Trust.

[a] In MBD.
[b] In $ billions for selected countries.

Saudi Arabia was able to exert great influence as a price moderator because of the volume of its oil exports and because of the position of Arabian Light as the OPEC marker crude. Also, from 1973 to 1979, when Saudi Arabia's influence was greatest, there was a continual threat to the OPEC price structure in the form of what was presumed to be a huge amount of excess capacity, already or potentially available to the Saudi government. Saudi oil production of 14-15 MBD would have cut Kuwait crude out of the market altogether, given its lower quality and higher price, in a pricing showdown scenario. Thus, the rest of OPEC was reluctant to push the Saudis too far while this threat hung so heavily.

The threat diminished in 1977, under the two-tier system which went into effect because of OPEC's inability to agree on a single price at the December 1976 Doha Conference (see note 51 in this chapter). Saudi excess production capacity looked nonexistent, as production declined generally for most OPEC members because of the drop in demand. Had Saudi Arabia been able *and willing* to put an extra 2 to 3 MBD of oil on the market, OPEC might have become a tame pet at Saudi command. But the two-tier system failed to make the other members of OPEC afraid of Saudi market power. The decline of Saudi Arabia as a power in OPEC became more and more evident in 1979-80 as it was virtually ignored by other OPEC members who raised their prices at will. The apparent demotion of Arabian Light to the status of just another Gulf crude is the official OPEC recognition of the new status quo.

Even if a subsequent OPEC Conference were to re-create a unified price structure for OPEC crudes, the resultant prices would surely be greater than Saudi preferred prices. The Saudis may occasionally be able to head a "Gulf coalition" at a second or lower price tier, but this is a far cry from controlling OPEC. The possibility that an Arab coalition might come about which could dominate OPEC price policies has become less likely since the line-up of Arab states on both sides of the Iran-Iraq conflict. OPEC itself seems to be losing its hold on oil prices, permitting member governments to assume greater pricing autonomy. The idea that "the Arabs" have controlled or do control OPEC oil prices seems more than ever another fiction of oil consumer governments.

OAPEC and OPEC

We have seen little evidence that the OAPEC members of OPEC control or exert an influence in OPEC disproportionate to their numbers or their oil exports. High OPEC oil prices have been maintained since 1973-74, not by Arabs, Saudi Arabia, or any OPEC coalition, but by them in concert with oil companies, consumer governments, and, most important, by non-OPEC oil producers, which have enjoyed being free riders on an OPEC-set global oil price floor.

The power of OPEC is a transitory phenomenon. OPEC as it is today will probably not outlive our children. Conceived as a means to exert collective pressure against the multinational oil companies, the organization has learned its lessons well and has become a sincere imitator of their techniques. OPEC's time is limited, however, because of declining global oil supplies and, perhaps, because there isn't much of a need anymore for an international effort to keep a floor beneath oil prices. "Is OPEC Redundant?" asks a British oil journal.[92] Perhaps.

OPEC's Arab members have had to take more than their share of the blame for the organization's success in helping its members to get control of the production and pricing of their oil. The OAPEC oil embargo may have been the key to OPEC's ability to hold up oil prices until it could get help from other quarters. Some of the biggest price hawks have been Arab OPEC members, but the biggest price dove is an Arab country too. Arabs have opposed Arabs on oil prices since December 1974. It would be inaccurate to attribute oil price increases to "the Arabs."

Saudi Arabia has occupied a special place in OPEC and in OAPEC. It has more oil than any other OPEC/OAPEC member and was crucial to the oil embargo decision. But the Saudi government's reluctance to go along with higher prices and production prorationing also made it a problem for the other members of OPEC. Saudi Arabia began to lose its power to threaten other OPEC members into agreeing with its positions in 1977 and was officially isolated in 1980 when its Arabian Light was demoted from its position as the only official marker crude. Regardless of whether Saudi Arabia may eventually lead a Gulf producers price coalition, its ability to set a price ceiling for the African crudes is gone.

OAPEC is dependent on OPEC. Financially, OPEC's decisions have enabled Arab oil producers to make enough money to support an Arab oil organization and its offspring: the joint venture companies. OPEC also serves as an alternative arena for making decisions on oil prices. As the Arab oil producers have disagreed so long and so much in this regard, keeping price quarrels out of OAPEC can only have added to its stability and, more important, to its longevity and increasing clout with member governments.

OPEC was dependent before 1979 on production restraint by its rich Arab members as a means of holding up oil prices. Since then, OPEC members have approved in principal the idea of limiting production in times of surplus oil supplies, although an actual prorationing program was vetoed at the May 1980 Taif meeting.[93] Voluntary limitations were set by Algeria in 1980, not one of the rich Arab countries, because much higher oil prices supported a production cut. The role of the rich Arab countries in limiting oil production may have been taken over unintentionally in 1980 by Iran, which raised its prices high enough to discourage purchase of Iranian crude. Iranian oil production in May was estimated at 1.5 MBD, lowering income drastically and forcing the nation to consider adopting an emergency budget.[94] While Iranian oil prices remain far out of line with other Gulf prices, Iran will suffer production cuts as the marginal producer. However, a return to a more "normal" set of price differentials will make production restraint by Arab countries important once again.

Summary

Since October 1973, OPEC, rather than the multinational oil companies, has acted as a petroleum cartel. OPEC sets prices rather than relying on production prorationing as the means to collect more of the economic rents associated with oil production.[95] Because all the relevant actors in price and production policymaking are OPEC members, and because the cartel sets a price rather than a production policy, the bargaining mode within OPEC is chiefly distributive. The price level considered as a pure public good is supported by the ability of OPEC members to view themselves as a community whose collective interest is best served by the agreed-upon price, regardless of whether the price is optimal for any member. It is also supported by its acceptance by non-OPEC oil producers acting as free riders on the OPEC-set price level.

OAPEC members have some potential power to control OPEC because, collectively, they are the major producing group, hold the majority of OPEC petroleum reserves, and can look forward to retaining that status throughout the period during which OPEC can set oil prices more or less at will. But political and economic divisions within OAPEC prevent its members from forming a single bloc to control OPEC policy for purposes specific to OAPEC members. OAPEC is not the petroleum cartel, even during times when OPEC is dependent on disproportionate production restraint on the part of OAPEC members to maintain the cartel price. The two organizations have been mutually dependent because of the contribution of OAPEC members to maintaining the cartel price and because of the probable inability of OAPEC itself to produce or implement a pricing policy as successfully as OPEC can.

Notes

1. F. M. Scherer, *Industrial Market Structure and Economic Performance* (Chicago, 1970), p. 158. A cartel is thus a subset of the general category "oligopoly," which refers to a market characterized by few sellers, each of whom "recognizes that its output decisions have a perceptible influence on price. . . . "(Ibid., p. 10). Oligopolistic markets are also characterized by significant barriers to the entry of new sellers (ibid.). A cartel is an overt collusive arrangement among the "few sellers." Oligopolies may also thrive under a system of "tacit collusion" among the sellers (see A. D. Neale, *The*

Anti-Trust Laws of the U.S.A., 2d ed. [Cambridge, Mass., 1970], Chap. III). OPEC appears to operate both overtly and tacitly when it sets prices.

2. Richard E. Caves, *International Cartels and Monopolies in International Trade,* Harvard Institute of Economic Research (Cambridge, Mass., 1977), p. 2.

3. Richard Schmalensee, "Resource Exploitation Theory and the Behavior of the Oil Cartel," *European Economic Review,* April 1976; also Fadhel Al-Chalabi, "Pricing of OPEC Crude Oil: A Case for the Valuation of Depletable Resources in Relation to Economic Development" mimeographed (Kuwait, 1977).

4. Among those who take this view are Paul Leo Eckbo, *The Future of World Oil* (Cambridge, Mass., 1975); and Øystein Noreng, *World Politics in the 1980's : Patterns of Cooperation* (New York, 1978); and Noreng, "Friends or Fellow Travelers? The Relationship of Non-OPEC Exporters with OPEC," *Journal of Energy and Development* 4 (Spring 1979).

5. See the discussion of the concept of a "just price" for oil in Charles F. Doran, *Myth, Oil and Politics* (New York, 1977), Chap. 1.

6. Eckbo, *The Future of World Oil,* pp. 76-77.

7. M. A. Adelman, *The World Petroleum Market* (Baltimore, 1972), pp. 76-77.

8. Paul W. MacAvoy, *The Economic Effects of Regulation* (Cambridge, Mass., 1965), pp. 21-22.

9. Eckbo, *The Future of World Oil,* p. 26.

10. See the OPEC Statute, Article 2.

11. *The Middle East,* 21 July 1976.

12. Eckbo, *The Future of World Oil,* Chap. 3. Thus OPEC is a cartel in the "loose constructionist" sense (see Chapter 1).

13. See Caves, "International Cartels," pp. 4-5.

14. Zuhayr Mikdashi, *The Community of Oil Exporting Countries* (Ithaca, 1972), Chap. 5.

15. "We, as the owners of the oil and the master of this land, must have a say in the production of this wealth because the needs of this country are clear. No firm, no company, no organization can tell us, merely because it has an agreement with us, that we will produce and export so much of your national wealth but you cannot touch the rest...." Speech by the Shah of Iran reported in *MEES,* 22 March 1968, p. 2a.

16. Interview with OPEC Deputy Secretary General Fadhel Al-Chalabi in Vienna, 10 June 1981.

17. Mikdashi, *The Community of Oil Exporting Countries,* pp. 114-15.

18. Ibid., p. 116.

19. This procedure was followed openly during the tedious negotiations on royalty expensing (treating royalties as a cost of production rather than as an exemption for income tax purposes) during the 1960s. At one point the companies offered a proposal that was viewed as the best obtainable by five of the six producers affected by the proposed agreement, with only one dissenting. The decision of the Seventh Conference was to "recognize the freedom of action of each member country in regard to the issue concerned." (Fuad Rouhani, *A History of O.P.E.C.* [New York, 1971], pp. 229-33, quote on page 233). The companies created an environment that fostered individualistic behavior by requiring each host country to negotiate individually with companies operating on its territory. As Libyan petroleum policy diverged from OPEC consensus positions after the coup in 1969, the companies denounced OPEC

leapfrogging, as gains made by Libya became the bargaining floor for the other members of OPEC wishing to catch up to the leader. See also J. E. Hartshorn, *Objectives of the Petroleum Exporting Countries* (Nicosia, Cyprus, 1978), pp. 4-8.

20. Robert Keohane and Joseph Nye, *Power and Interdependence* (Boston, 1977), Chap. 3. Actor independence has been cited by one authority, Erwin Hexner, as necessary for collusion to be cartel behavior (London, 1946).

21. Ibid.; also Richard N. Cooper, "Economic Interdependence and Foreign Policy in the 1970's," *World Politics* 24 (January 1972).

22. This occurs when nongovernmental actors make decisions to intervene in another nation's economy. Examples of such intervention are investment, disinvestment, and movement of industries for tax avoidance or to get cheaper labor. For a detailed discussion of international economic sensitivity, see Cooper, "Economic Interdependence and Foreign Policy in the 1970's"; Cooper, *The Economics of Interdependence* (New York, 1968), Chap. 1; and Keohane and Nye, *Power and Interdependence*, Chap. 2.

23. For example, there is amazing tolerance within OPEC for overproduction by Saudi Arabia. However, it does not stop other members from cutting their own production to counter the Saudi action.

24. Arab oil exporters might modify that statement. They feel that private Norwegian producers have frequently underbid higher-cost OPEC producers and depressed prices. Noreng, "Fellow Travelers," p. 329.

25. Libya's situation vis-à-vis the oil companies was different from that of the other major Middle Eastern producers for two reasons: Libya granted many small concessions to many different companies (and combines) instead of a few large ones to one or two companies or combines; and it hosted several "independent" (non-integrated) companies for which it was the sole source or most important source of crude oil. The government could force production cuts upon dependent companies, as it did to Occidental Petroleum, which was forced to comply with its demands in order to survive. Particularly after proposals for a "safety net," or guarantee of dependent producers' crude needs by companies with crude sources outside Libya, fell through, dependent "independents" like Occidental faced the dilemma of either giving in or going out of business. Of course, they gave in. (See the testimony of George Schuler in U.S., Congress, Senate, Subcommittee of the Committee on Foreign Relations, *Multinational Corporations and U.S. Foreign Policy, Hearings*, 93d Cong., 1st Sess., Part 5, 1974 (hereafter cited as *Hearings*); and Christopher Rand, *Making Democracy Safe for Oil* (Boston, 1975), Chap. 12.

26. Ibid., also Richard C. Weisberg, *The Politics of Crude Oil Pricing in the Middle East, 1970-1975* (Berkeley, 1977), pp. 73-74.

27. Ibid., Chap. 5.

28. George Lenczowski, "The Oil Producers," in Raymond Vernon, ed., *The Oil Crisis* (New York, 1976), p. 68.

29 .See Farouk A. Sankari, "The Character and Impact of Arab Oil Embargoes," in Naiem Sherbiny and Mark Tessler, eds., *Arab Oil: Impact on the Arab Countries and Global Implications* (New York, 1976).

30. Charles F. Doran, "Conflict, Cohesion and Coalition Formation in OPEC: A Quantitative Assessment" (paper presented at the 1977 Annual Meeting of the American Political Science Association, 1-4 September 1977 in Washington, D.C.), pp.15-16.

31. Doran, "Conflict, Cohesion and Coalition Formation in OPEC"; see also Figure 2.

32. See M. F. Adelman, "Is the Oil Shortage Real? Oil Companies as OPEC Tax Collectors," *Foreign Policy* 9 (Winter 1972-73). Another reason why the price structure did not collapse might be that the unshaved prices were actually out of line with the rest of OPEC prices.

33. There is great disagreement on the role of oil prices in inflation. A 1976 Brookings Institution book coauthored by Charles Schultz put the OPEC contribution from the 1973-74 price hikes at about 1½ percent. The June 1980 Venice Summit Communiqué of eight OECD heads of state blamed the majority of the inflation experienced by oil-importing countries on OPEC. OPEC Secretary General Rene Ortiz immediately issued his own statement taking issue with the Venice Summit Communiqué. There is general agreement that prices for manufactured goods from the OECD have increased at a higher rate than oil prices. See Walter J. Levy, "The Years That the Locust Hath Eaten."

34. *Middle East Economic Survey (MEES)*, 30 June 1980, p. 8.

35. *Quarterly Report*, June 1980, reported in *MEES*, 30 June 1980, p. 9.

36. Noreng, "Fellow Travelers," p. 319.

37. John Bloom, "Profits? What Profits?" *Texas Monthly*, July 1980.

38. Scherer, *Industrial Market Structure*, p. 283.

39. A dissenting view to this reasoning regarding the rationale for and the success of OPEC's oil price policies is that of Ali D. Johany. Johany's view is that OPEC is not a cartel and couldn't possibly form a cartel because its members have too many political and economic differences ever to adopt a single production plan. As explained in Chapter 1, this is not an uncommon view for scholars from the Third World, and is consistent with a large body of literature on OPEC written from that perspective.

40. Weisberg, *The Politics of Crude Oil Pricing*, p. 147.

41. Theodore J. Lowi, "Four Systems of Policy, Politics and Choice," *Public Administration Review* 32 (July/August 1972), 299.

42. Theodore J. Lowi, "American Business, Public Policy, Case Studies and Political Power," *World Politics* 16 (July 1964), 713.

43. Lowi regards the legitimate use of coercion as the distinguishing hallmark of government and types policies according to how the coercion is applied to achieve desired results. Wilson's views on policy are closer to those represented here. He feels that the Lowi categories are difficult to operationalize and do not provide any indication of how the arenas form around policy issues. Wilson uses the differential incidence of policy costs and benefits to type policies. Comparing the policies that each author feels epitomizes his policy categories we can see that the typologies do not differentiate among policies in the same way (see Figure 6). Wilson combines Lowi's redistributive and constituent categories while dividing the Lowi regulation into two. See James Q. Wilson, *Political Organizations* (New York, 1973), Chap. 16.

44. Ibid., pp. 333, 335.

45. An example often cited in the U.S. domestic policy context is the National Labor Relations Board (NLRB). This body regulates relations between employers and employees, business and unions, but, again, the ultimate bearer of the costs of agreements between these two, the consumer, is excluded from representation before the NLRB. See Wilson, *Political Organizations*, p. 335, and also Roger Noll, *Reforming Regulation* (Washington, D.C. 1971), pp. 48-51.

Figure 6: *Public Policy According to Lowi and Wilson*

A. Lowi   Likelihood of Coercion

APPLICABILITY OF COERCION

	Remote	Immediate
Individual Conduct	DISTRIBUTIVE Farm Subsidies Land Grants	REGULATIVE Regulatory Commission Policies
Environment of Conduct	CONSTITUENT National Defense	REDISTRIBUTIVE Income Tax

B. Wilson   Costs

BENEFITS

	Distributed	Concentrated
Distributed	National Defense Social Security Interstate Highways	FDA Regulation Auto Safety Regulation
Concentrated	Veterans' Benefits Rural Electrification	National Labor Relations Board Interstate Commerce Commission

Source: Adapted from Theodore J. Lowi, "Four Systems of Policy, Politics and Choice"; and from James Q. Wilson, *Political Organizations*.

46. See Chapter 1.

47. Lowi, "American Business."

48. An exclusive benefit is one to which access can be denied feasibly, such as passage over a bridge, while the cost of permitting such access to an additional individual is essentially zero.

49. Even minimal contributions from the less advantaged reduce the cost of provision to the more advantaged, who would have paid that much more to provide the benefit for themselves alone.

50. It might be argued that the market price for oil is the result of conflict over "high" prices and "higher" prices reflecting the respective preferences of high-reserves and low-reserves nations and thus is not a "community of interests" type of policy. I argue that it is a policy of this type because it is impossible to know what the true optimal price really is. The "high" price and "higher"price preferences are not strictly rival but are rather "worst case" projections by individuals jockeying for leadership status within OPEC and a hedge against these worst case scenarios. Because there is no way to show that either the "high" price or the "higher"price will, in the long run, maximize joint cartel profits, the arguments cannot really be called redistributive bargaining.

51. Figure 7 shows monthly barrel-per-day production rates for Saudi Arabia and the major producers among the high price preference countries for the last half of 1976 and the first half of 1977 when the formal two-tier pricing system was in effect, and into the immediate post two-tier era. The Saudis were estimated by industry analysts (Walter J. Levy SA) to have had a production capacity of at least 12 million barrels per day in 1977, a level of production never reached during the period of the two-tier system, even before the destructive fire in the Abqaiq field in May. Saudi Arabia, Iran, and Iraq experienced sharp drops in production in January, following higher than normal production levels in the last quarter of 1976 (probably caused by company anticipation of a price rise in the new year). But during the entire period of the two-tier regime, the Saudis never produced at capacity, an action which might have depressed oil prices toward the Saudi preference level. A difference of means test comparing the average Saudi production rate during the two-tier system with their average rate during the other ten months reported in Figure 7 is not significant at the .05 level. The data is from *Petroleum Economist (PE)*, December 1977, p. 503; the estimates of production capacity are from *Oil and Gas Journal (OGJ)*, 16 August 1976, p. 52.

52. Lawrence R. Raicht, "The Politics of Energy" (talk given at the Department of State Conference on U.S. national interests in the Middle East, 15 December 1977, in Houston, Texas).

53. Saudi Arabia had been threatening, at least since the June 1974 OPEC Conference meeting, to leave OPEC because of its disagreement with OPEC oil-pricing policy. In practice, its disagreement was expressed by Saudi maintenance of oil prices at levels they preferred. Differential pricing was common by September 1974 throughout OPEC, when an agreement to halt price increases was reached, apparently in an effort to persuade low-price members to bring their prices into line with the rest of the members of OPEC.

However, continued disagreement made differential pricing endemic throughout OPEC during 1975 and 1976. The 15 December 1976 Doha Conference marked the initiation of a formal two-tier pricing system beginning the following month. The situation, in fact, was little different from what had existed for the two previous years. Indeed, Venezuela only raised prices on some kinds of crude. Indonesia raised prices less than the agreed upon 10 percent, while Iran, Kuwait, Qatar, and Libya raised their prices more than 10 percent (see *PE*, July 1977, p. 254).

The formal two-tier system was more an instance of public disagreement than a different pricing regime. The resolution of price differentials that took place after the end of the formal two-tier regime appears to have been made individually and for marketing reasons rather than as a result of a "return" to identical prices for similar

Figure 7: *Oil Production in Selected OPEC Countries During Two-Tier System in 1977*

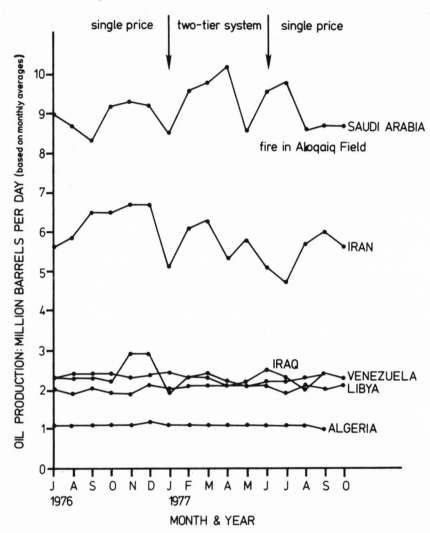

crudes (*The Houston Post*, 16 September 1977, p. E2). Throughout the 1974-78 period, there seems to have existed a tolerance of variation from the target price set or implied by OPEC Conference resolutions. Individual pricing variations have been used to exert pressure on other OPEC members with regard to future policy decisions, but at no time did a price moderate engage in action that might have threatened the general price structure. Both sides to the pricing controversy practiced differential pricing. Differential pricing was finally institutionalized within limits at the March 1979 Extraordinary Meeting of the OPEC Conference.

54. The amount of reserves a country has influences its status in bargaining for increased production development activities by oil companies, in its political relationships with consumer countries and in bargaining relationships within the cartel itself. The advantage to the oil-exporting country in any of these situations varies inversely with reserves levels.

55. *OGJ*, 31 March 1980, p. 70.

56. *OGJ*, 7 April 1980, p. 50.

57. Exercisers of production restraint face strong internal and external demands to increase production. Internal demands emphasize income gains to be made from greater production, while external demands have their roots in consumers' desires to ensure supply to meet current demand plus, possibly, a desire to weaken prices. However, in the case of Kuwait, these demands are opposed by others that stress the need to conserve the nation's only natural resource as long as possible, perhaps for future use by native industries not yet existing. Venezuela wants to preserve its present position as a producer, as well as to conserve fuel resources for its own eventual use. Production.policy at any time represents a compromise among conflicting demands on individual countries.

58. Some type of attribute analysis of the oil-producing countries has been done as a part of several of the major studies of the industry. Examples can be found in Adelman, Eckbo, and Doran, *Myths*, all previously cited, and in Nazli Choucri, *The International Politics of Energy Interdependence* (Lexington, Mass., 1976).

59. Doran, "Conflict, Cohesion."

60. Ibid., Table 5. Doran feels that the general level of turmoil and disruption of normal international transactions produced the anomalous results he obtained for 1973.

61. Ali Khalifa Al-Kuwari, *Oil Revenue in the Gulf Emirates* (Boulder, Colorado, 1978) Chap. 2; and David Crawford, *Getting Work in OPEC* (London, 1975), p. 45.

62. Libya is different for the reasons cited in note 25. Its petroleum laws encouraged prompt development of discovered fields, and its ability to play the interests of some producers, particularly the independents, off the interests of the rest, has given Libya much more political and economic leverage over its corporate guests than the other Middle Eastern oil producers had (see Christopher T. Rand, *Making Democracy Safe for Oil* [Boston, 1975], Chap. 12).

63. By economically indifferent I mean that the advantages versus the disadvantages of a "high" price as opposed to a "higher" price for oil are about equally balanced for these two nations.

64. Noreng, *World Oil Politics in the 1980's*, pp. 78-79. Paul Eckbo also used economic dimensions to categorize OPEC members and got coalitions very similar to Doran's (*The Future of World Oil*, pp. 80-82).

65. Ibid., pp. 80-81.

66. Eckbo, *The Future of World Oil*, pp. 80-82.

67. Iraq is underexplored.

68. Central Intelligence Agency, "The International Energy Situation: Outlook to 1985" (Washington, D.C., April 1977).

69. Iranian monthly production totals for October 1978 were as follows (in millions of barrels):

1978:	October	171.8
	November	104.8
	December	73.5
1979	January	13.8
	February	19.6
	March	72.9
	April	108.0

Petroleum Economist (*PE*), April 1980, p. 181.

70. *OGJ*, 28 April 1980, pp. 47-48.

71. *MEES*, 3 April 1978, p. 1.

72. *MEES*, 1 January 1979, pp. 6-7.

73. Quoted in *MEES*, 1 January 1979, p. 6.

74. Ibid.

75. *MEES*, 12 February 1979, p. 1; *MEES*, 29 January 1979, pp. 2-3; 26 February 1979, p. 1; *Supplement*, 27 February 1979, p. 1.

76. *MEES*, 5 March 1979, pp. 9-10.

77. "The Need for Flexibility in the OPEC Pricing System," *OAPEC Bulletin*, March 1979, pp. 13-17.

78. *MEES*, 2 April 1979, pp. 1-5.

79. *MEES*, 11 June 1979, p. 1.

80. *MEES*, 2 July 1979, pp. 1-4.

81. *MEES*, 9 July 1979, p. 1.

82. *MEES*, 15 October 1979, pp. 1-2.

83. *MEES*, 29 October 1979, p. 1.

84. *PE*, February 1980, p. 47; *MEES Supplement*, 24 December 1979, p. 2; John R. Emshwiller and James Tanner, "OPEC Power Struggle to Have Usual Result: Further Price Increases," *Wall Street Journal*, 6 February 1980.

85. *MEES Supplement*, 24 December 1979, p. 2.

86. *MEES*, 29 January 1979, pp. 2-3.

87. *Oil and Energy Trends*, 18 January 1980, pp. 2, 11.

88. Editorial by Fadhil Al-Chalabi, editor-in-chief of the *OPEC Review*, Summer 1980, pp. i-iv.

89. Ian Seymour in *MEES*, 6 June 1980, pp. 1-5.

90. Chalabi, "Editorial," p. ii.

91. Ibid., pp. ii and iii.

92. This is a judgment of the *Petroleum Economist*, which published an article in January 1980 saying that OPEC had "lost control" of oil prices. If one compares the 1978-79 shortage to the one of 1973-74, there seems to be support for this position.

93. *MEES*, 12 May 1980, p. 3.

94. *MEES*, 9 June 1980, p. 1.

95. It had been only marginally successful in catching up to the European governments' share of the larger and larger economic rent from oil prior to 1979 (see Table 27).

TABLE 27 *Distribution of the Rent from Oil*

	1961	1973	1975	1973/1975
Producer	$0.76	$2.30	$10.10	$7.80
Revenue	6%	11%	30%	339%
European	$7.10	$11.40	$14.90	$3.50
Taxes	52%	56%	45%	31%
Company	$5.70	$6.80	$8.20	$1.40
Margins	42%	33%	25%	21%
Weighted	$13.60	$20.50	$33.20	$12.70
Average				62%

Source: Ali M. Jaidah, "Pricing of Oil: Role of Controlling Power," *OPEC Review* 1 (June 1977), 15.

Chapter 6

Prospects for the Eighties

A diagram of the varying power of the Organization of Arab Petroleum Exporting Countries might resemble a somewhat distorted sine curve. The curve would begin in 1968 almost at the peak of a cycle and then plunge down in 1971-72, during the time of the membership crisis. The trough then would have been passed, although the curve would remain below the zero line until 1977. Then it would continue upward toward a new peak through 1980. Another way to put this assessment is to say that OAPEC's prospects have looked very good since May 1978, nearly as good as they promised to be in the organization's early days. In two of the three general areas discussed in this work, OAPEC has advanced. The goals of economic development and oil policy coordination are currently being pursued with new vigor. However, OAPEC's influence in OPEC has diminished from its peak during the Arab oil embargo.

Economic Development and Technology Transfer

OAPEC's role in economic development and technology transfer has been recently enhanced by the creation of the Arab Engineering Consulting Company (AECC), whose establishment was announced at the June 1980 meeting of the OAPEC Council of Ministers. The AECC is to supply a range of engineering services to its customers, some of which will include OAPEC and its joint venture companies and others perhaps the member country national oil companies (NOCs), which are shareholders in the new venture. Participation by other than OAPEC member governments is a significant departure from the usual pattern of shareholding in an OAPEC company. Participation by the NOCs gives them a stake in the new venture's success. It expands the levels of government shares in liabilities and profits, by allocating portions specifically to the NOCs. Allowing the national oil companies to participate directly in the fortunes of the AECC may be a greater incentive to the NOCs to support the infant venture than participation by governments alone has elicited for other OAPEC joint venture companies.

The establishment of the new joint venture company illustrates OAPEC's continuing commitment to technology transfer. As I pointed out earlier, Arabs, like other economically developing peoples, are reluctant to employ high-level technical personnel from their own countries. Developing-country preferences for scientists and engineers from developed countries are probably linked more closely to a quest for international prestige than to a judgment on the qualifications of native workers. Regardless of their source, such preferences definitely retard progression toward economic self-sufficiency, which most developing-country governments say they support. OAPEC has always been committed to the training and subsequent employment of Arab workers at all levels of the petroleum industry, even when such a commitment has been very costly. The OAPEC imprimatur on a high technology operation like the engineering consulting firm is additional evidence of the continuing interest of the organization in achieving Arab technology independence.

Arab Community

Institutional community within OAPEC is both weaker and stronger in 1980 than it was in 1977. On the one hand, the suspension of Egypt is a deliberate tearing of the fabric of community by ostracizing a member. Even though Egypt only marginally fits the amended charter description of an OAPEC member as a country that depends on oil exports for a significant percentage of its national income, Egypt's oil dependence is growing and its vulnerability to a disruption of its foreign oil sales is at least as great as Bahrain's or Syria's. However, the whole history of OAPEC is inextricably tangled in the threads of the Arab-Israeli conflict. It should, perhaps, be less surprising that OAPEC has suspended Egypt than that other organizations from the Islamic Conference to the General Federation of Arab Chambers of Commerce, Industry and Agriculture have done so.

On the other hand, Arab community measured as member cooperation and oil policy coordination has grown. Although OAPEC exists to coordinate member petroleum policy, this task has been made difficult by some members that have been reluctant to divulge information about their domestic oil industries and others that have been slow to support some of the OAPEC projects. That many of these and other difficulties have been resolved is due in large part to the political sophistication of the current secretary general of OAPEC, Ali Attiga. Attiga's decision to "go public" and expose in a general way the nationalistic behavior of OAPEC members seems to have been the turning point in OAPEC's fortunes in the 1970s. Since then, a new joint venture company, a new subsidiary of an already existing company and an OAPEC Mediterranean dry dock have been initiated. The protocol of the OAPEC Judicial Board was finally written and ratified, and support voted and paid to the OAPEC tanker company, AMPTC, to help it over its financial problems. The long-standing disagreement between Iraq and Saudi Arabia over the siting of the Arab Petroleum Training Institute was also resolved. Most of these issues were politically sensitive and potentially divisive. That they have been taken care of with a minimum of acrimony and dissention within the organization is a tribute to the secretary general as well as to the members.

A recent difficulty regarding member payments to AMPTC will illustrate the growing organizational maturity of OAPEC. Iraq and Algeria had refused to pay their shares of the member assessments to cover AMPTC's losses because of their dissatisfaction with the company's management. At their June 1980 council meeting, in a masterly compromise, the oil ministers decided that Iraq and Algeria must pay their shares and that a special committee would be set up to investigate the company's management and to suggest improvements. The principal of collective decisionmaking was upheld, while the rights of shareholders to disagree with the management of the joint venture companies was also established.

The ratification of the Judicial Board protocol is perhaps the most telling piece of evidence for OAPEC's institutional strength. At this writing, the OAPEC oil ministers have agreed that judges will be appointed at the next scheduled council meeting. The as yet nonexistent Judicial Board is far from any attempt to exercise its authority but it is remarkable that such an institution was ever agreed to since the 1969 coup in Libya made political and economic differences within OAPEC a constant fact of life.

OAPEC in OPEC

We have seen that while OAPEC's dependence on OPEC to set oil prices has been explicit, the dependence of OPEC on OAPEC has been indirect. Prior to 1979, Arab countries were often the OPEC members most willing to absorb oil production cuts to sustain OPEC prices. The disorder in the global oil market since then makes it difficult to say whether the Arabs or anyone else controls or exerts a deciding influence on OPEC-set oil prices. During this period OPEC prices have been set by individual governments rather than by the OPEC Conference. The coalitions which have emerged since 1978 have tended to pit the African producers and Iran against the other members. Despite the numerical superiority of the moderate price camp, oil prices have risen sharply. The OPEC Conference has also acted to curb the moderating influence of Saudi Arabia on oil prices by adopting a theoretical marker crude thereby cutting loose Arabian Light from its former position as the anchor of the OPEC price schedule.

Since January 1979, collective control over oil prices by OPEC has been diminished by the organization's inability to decree an official schedule for differentials and its unwillingness to tackle the issue of contract premiums. Thus, even when some sort of OPEC oil price floor and/or ceiling has been adopted, the actual selling prices of the various member crudes have only sometimes conformed to the collective pricing decision. The prices have instead been functions of national governments' decisions as to what their oils' quality and locations rated within the official price range and how much extra each could charge in addition to that per barrel price as a premium for selling any oil at all. Algeria, for example, charges an exploration fee per barrel of exported oil and Kuwait charges premiums on new contracts. If such extra charges are not somehow made to conform to a general price schedule, OPEC may be out of business as the global oil price setter.

Although OAPEC cannot be shown to have had an appreciable effect on oil prices since the end of the Arab oil embargo in 1974, OAPEC cooperates with OPEC in other ways. OAPEC has tried, without much success so far, to act as the vehicle for direct long-term oil producer-oil consumer negotiations. Its "opportunities for cooperation" seminars, cosponsored by various European countries, have been moves in that direction, and, ironically, the most productive liaison so far has been between Scandinavian oil producers and their Arab counterparts.

Another way in which OAPEC is important to OPEC is in its role as an agent of policy coordination and economic development for its members. Occasionally, one finds mention of eventual OPEC joint venture corporations along the OAPEC model in the writings of OPEC officials. OPEC joint ventures would be difficult to set up because the largest financial surpluses in OPEC are held by Arab countries. Non-Arab members of OPEC might be reluctant to invest in an Arab-dominated venture and the Arabs might prefer to confine their joint venture participation to other Arab countries.

Another service performed, however unwillingly, by OAPEC for OPEC is to be the scapegoat bearing the brunt of consumer complaints about high oil prices and uncertain oil supplies. Since December 1973, when the Shah of Iran acted openly as the biggest price hawk in OPEC and Saudi Arabia the greatest dove, "the Arabs" have been blamed for almost every real or imagined wrong committed by oil exporters against oil consumers. Perhaps with the producing governments acting as individuals in oil pricing, the onus for high oil prices may shift to Iran or the African producers, those charging the highest prices. But it is not likely that such a shift will proceed very far in the United States, regardless of its progress in other oil-consuming states.

Conclusions

The Organization of Arab Petroleum Exporting Countries is in a strong position. It has made great strides toward developing unique international agencies which function at some cost to the national autonomy of OAPEC members. In spite of their disenchantment with OAPEC and its offspring during most of the 1970s, the member governments eventually decided to stand behind what they had created and to expand and strengthen the organization.

OAPEC's efforts to promote regional economic development and technology transfer may prove to be its most enduring and important contribution to the Middle East region. Year after year, OAPEC trainees are turned out by instruction programs supported by the joint venture companies and by the OAPEC Secretariat. The Arab Petroleum Training Institute will eventually take its place as a regional research institution as well as a teaching academy. OAPEC also encourages scientific activity through its seminars and conferences and supports in-house research into various aspects of the petroleum industry.

Even though OAPEC follows OPEC's oil price decisions rather than commanding them, it is a leader in developing ways to become indispensible to its members. If or when OPEC's control of member oil prices disappears altogether, it will have to devise new reasons for continuing to exist. The OPEC Development Fund, which channels some of the multilateral foreign economic assistance disbursed by oil-exporting countries, is one reason for OPEC to continue to function, but it is not a sufficient reason. There is little incentive to maintain an organization just to give away one's money, particularly when there are similar organizations, more tightly tied to member government policies than OPEC is, to do this. OPEC seems to be bidding for a role as a buffer in negotiations between the rich and poor nations of the world. It may be uniquely qualified to do this because it has things both sides want: oil and money, which it can use to coax cooperation from the world's economic combatants. But the achievement by OPEC of new roles and functions is still some distance away.

OAPEC is continually diversifying, entering more and more areas that now fall under the heading of petroleum industry but that are evolving in the more general direction of economic development per se. Institutionally, OAPEC is one of the most interesting international organizations created since the end of World War II. It is remarkably autonomous given its membership and its early trials. OAPEC is also one of the two or three really innovative international organizations that act as models for others to emulate. The 1980s promise to be OAPEC's decade. After thirteen turbulent years, the organization is strong and enjoys its member's support. One cannot but hope that OAPEC will prosper as a model for economic development and international cooperation.

Agreement of the Organization of Arab Petroleum Exporting Countries*

In the Name of God, The Gracious, The Merciful

This Agreement entered into the city of Beirut on 9 January 1968, corresponding to 9 Shawwal 1387 between and among:

> The Government of the Kingdom of Saudi Arabia, represented by His Excellency Sheikh Ahmed Zaki Yamani, Minister of Petroleum and Mineral Resources;

> The Government of the State of Kuwait, represented by His Excellency Mr. 'Abd al-Rahman Salim al-'Atiqi, Minister of Finance and Oil;

> The Government of the Kingdom of Libya, represented by His Excellency Mr. Khalifah Mousa, Minister of Petroleum Affairs.

Agreement for the Establishment of an Arab Organization for the Petroleum Exporting Countries

The Arab Petroleum Exporting Countries signatories to this agreement,

Recognizing the role of Petroleum as a principal and basic source of their income which they should develop and safeguard in such a way as to provide them with the utmost legitimate benefits, and

Being Aware that Petroleum is a depleting resource and that fact places upon them the charge vis-à-vis future generations of conserving it and the responsibility of utilizing the wealth derived from it in economically diversified investments in productive and developing projects having the essentials of vitality and prosperity, and

Believing that the rational utilization of this asset is related to the role of Petroleum in serving the economies of the consumer countries and consequently entails due consideration for the legitimate interests of said countries in obtaining Petroleum supplies for their markets on equitable terms conducive to the well-being of humanity, and

Partaking in the development and prosperity of the world Petroleum industry, and

*Reprinted with permission of OAPEC.

Anticipating the achievement of close and fruitful cooperation among themselves in such fields,

Have agreed upon the following:

A. *The Organization, Its Objectives and Powers*

Article ONE
(a) There shall be established in accordance with this Agreement an organization called "The Organization of Arab Petroleum Exporting Countries" hereinafter referred to as "the Organization."
(b) The Organization shall be domiciled in the City of Kuwait in the State of Kuwait.

Article TWO
The principal objective of the Organization is the cooperation of the members in various forms of economic activity in the petroleum industry, the realization of the closest ties among them in this field, the determination of ways and means of safeguarding the legitimate interests of its members in this industry, individually and collectively, the unification of efforts to ensure the flow of petroleum to its consumption markets on equitable and reasonable terms, and the creation of suitable climate for the capital and expertise invested in the petroleum industry in the member countries.

In pursuit of the said objectives, the Organization shall in particular:

(a) Take adequate measures for the coordination of the petroleum economic policies of its members.
(b) Take adequate measures for the harmonization of the legal systems in force in the member countries to the extent necessary to enable the Organization to carry out its activity.
(c) Assist members to exchange information and expertise and provide training and employment opportunities for citizens of member countries in members' countries where such possibilities exist.
(d) Promote cooperation among members in working out solutions to problems facing them in the petroleum industry.
(e) Utilize the member resources and common potentialities in establishing joint projects in various phases of the petroleum industry such as may be undertaken by all the members or those of them that may be interested in such projects.

Article THREE
The provisions of this Agreement shall not be deemed to affect those of the agreement of the Organization of Petroleum Exporting Countries (OPEC), and especially in so far as the rights and obligations of OPEC members in respect of that organization are concerned.

The parties to this Agreement shall be bound by the ratified resolutions of OPEC, and shall abide by them even if they are not members of OPEC.

Article FOUR

(a) The Organization shall enjoy a juridical personality and the legal capacity entitling it to exercise in the territory of each and every member all the powers of juridical persons.

In particular, the Organization may acquire and own moveable and immoveable property, perform legal transactions, as well as sue and be sued in its own name.

(b) The Organization shall enjoy in the territories of its members such immunities and privileges as are necessary for the fulfillment of its objectives and the performance of its activities, and the premises it occupies shall enjoy immunity. All such shall be defined in detail in a protocol to be appended to this Agreement.

Article FIVE

The Organization may conclude agreements with members, or with other countries, or with a federation of states or with an international organization, and especially agreements for establishing joint projects in various phases of economic activity in the petroleum industry.

Article SIX

(a) The contractual liability of the Organization shall be governed by the law of the contract entered into. The liability for tort shall be governed by the common general principles of law of the members.

(b) The rights and duties of the Organization's personnel shall be subject to the provisions of this Agreement and to the statutes, regulations and decisions issued in accordance with it.

B. Membership of the Organization

Article SEVEN

(a) The founding members of the Organization are the signatories to this Agreement.

(b) Any Arab country may join the membership of the Organization provided the following conditions are fulfilled:

(1) Petroleum should constitute a significant* source of its national income.

(2) It should adhere to the provisions of this Agreement and amendments which may be made to it.

(3) The Council referred to in Article Eight below should approve its membership by a majority of three-quarters of the votes provided all the votes of the founding members are included.

*As amended.

C. Organs of the Organization

Article EIGHT

The Organization shall carry out its functions and responsibilities through the following organs:

First: A Council of Ministers, hereinafter referred to as "the Council."
Second: An Executive Bureau, hereinafter referred to as "the Bureau."
Third: A Secretariat General, hereinafter referred to as "the Secretariat."
Fourth: A Judicial Board, hereinafter referred to as "the Board."

First: The Council

Article NINE

The Council shall be composed of one representative from each member country, who shall be the minister of petroleum or someone enjoying a comparable degree of responsibility for petroleum affairs.

The representative may be accompanied by a number of assistants at meetings of the Council.

Article TEN

The Council is the supreme authority of the Organization, responsible for drawing up its general policy, directing its activity and laying down the rules governing it.

Without prejudice to the generality of the foregoing paragraph of this article, the Council shall be concerned with the following:

(a) Deciding on applications for joining the membership of the Organization and approving invitations to petroleum exporting countries to attend its meetings.

(b) Taking resolutions, making recommendations and giving advice with regard to the general policy of the Organization, or its attitude or the attitude of a particular member or particular members vis-à-vis a particular situation or issue or particular situations or issues.

(c) Choosing negotiators and appointing committees to negotiate on behalf of the Organization.

(d) Approving draft agreements reached by the Organization.

(e) Issuing the necessary regulations and amendments thereto.

(f) Approving the Organization's draft annual budget, and ratifying the end-of-year accounts.

(g) Appointing the Secretary General and Assistant Secretaries.

(h) Supervising and directing the work of the Bureau and the Secretariat.

(i) Matters which this Agreement or the regulations did not expressly provide for to be within the competence of any other organ.

Article ELEVEN

Voting in the Council shall be subject to the following provisions:

(a) Each member of the Organization shall have one vote.
(b) In order to constitute a quorum for the meetings of the Council, three-quarters of the total members must be present provided there shall be included among them two founding members at least.
(c) Statutes shall be issued by the Council, and resolutions on substantive matters shall require a majority of three-quarters of the total votes of the members, including those of at least two founding members.
(d) Resolutions of the Council on procedural materials shall be carried and its recommendations and advice adopted by a simple majority vote of all the members.
(e) The Council shall determine, by a simple majority vote of all the members, which matters shall be considered substantive and which procedural.

Article TWELVE

(a) Statutes shall be generally applicable and binding among all the members.
(b) Resolutions shall have binding authority on those whom such resolutions concern.
(c) Neither the recommendations nor the advice of the Council shall be of a binding nature.
(d) Without contravening the provisions of Article Twenty Three of this Agreement, a resolution which is binding on any one of the members shall provide for the solidarity of all the other members with the member concerned and shall, upon the request of such member, be accompanied by guarantees from the Organization for protection against any damage to which such member may be exposed and, should the occasion arise, for the sharing by all the members of the burden of such damage.
(e) Statutes or resolutions of a binding nature which are issued by the Council shall be subject to ratification by the competent authorities in the member countries according to the legal rules in force.

Article THIRTEEN

(a) The Council shall convene at the Headquarters of the Organization, and may also be convened in the territory of any member or of any other country if it deems it necessary.
(b) The Council shall convene at least twice a year; extraordinary sessions may be convened at the request of one of the members or the Secretary General.
(c) Representatives of the member countries shall assume the chairmanship of the Council by rotation according to the alphabetical order of the countries they represent, each for a period of one year.

Second: The Bureau

Article FOURTEEN

The Bureau shall be composed of one representative from each of the member countries, such representatives to be appointed by the country concerned. The representatives shall assume chairmanship of the Bureau by rotation according to the alphabetical order of the member countries each for a period of one year.

Article FIFTEEN

The Bureau shall have the following competencies:
(a) Consider matters relating to the application of this Agreement and the Organization's performance of its activities and functions.
(b) Submitting such recommendations and proposals as it sees fit to the Council concerning matters falling within the scope of this Agreement.
(c) Approve the staff regulations of the Secretariat and introduce appropriate amendments thereto after consultation with the Secretary General.
(d) Review the Organization's draft annual budget and refer it to the Council together with the Secretary General's observations.
(e) Draw up agendas for the Council.
(f) Such functions and tasks as may be assigned to it by the Council.

Article SIXTEEN

(a) The Bureau shall convene at the invitation of its chairman before every Council meeting in time to allow it to prepare the Council agenda.
(b) The Bureau may hold two meetings at other times at the invitation of its chairman, upon the request of one of the members or the Secretary General.
(c) The Bureau shall convene at the Headquarters of the Organization, and may also be convened in the territory of any member country or any other country if it deems it convenient.
(d) Resolution of the Bureau shall be carried by a majority of two-thirds of the votes of all members in accordance with the provisions of Paragraph (a) of Article Eleven of this Agreement.

Third: The Secretariat

Article SEVENTEEN

(a) The Secretariat shall be composed of the departments and committees laid down by the statutes and shall assume the planning, administrative and executive aspects of the Organization's activity in accordance with the statutes and directives of the Council.

(b) The seat of the Secretariat shall be the Headquarters of the Organization.

Article EIGHTEEN

(a) The Administration of the Secretariat shall be assumed by a Secretary General, aided by Assistant Secretaries who shall number no more than three unless the Council approves an increase in this number.

(b) The Secretary General and the Assistant Secretaries shall be nationals of the member countries, with adequate experience in petroleum affairs, and not more than two of them shall be selected from the same nationality.

(c) The Secretary General shall be appointed by decision of the Council for a period of three years extendable for a further period or periods.

     The Assistant Secretaries shall be appointed by decision of the Council for a period of four years extendable for a further period or periods.

(d) The Secretary General shall be the official spokesman for the Organization. He shall be its legal representative within the limits of the powers vested in him by this Agreement, the statutes, resolutions and directives which may be issued in accordance with it. The Secretary General shall be responsible before the Council for carrying out the tasks and duties of his office.

Article NINETEEN

(a) The Secretary General shall be charged with administering and directing the Secretariat, effectively supervising all aspects of its activities, and carrying out the tasks assigned to him by the Council.

     The departments and committees of the Secretariat shall carry out their tasks in accordance with his instructions and directives within the limits of the powers vested in him.

(b) The Secretary General shall carry out the duties of Secretary to the Council and the Bureau whenever either of them is meeting. He may delegate one of the Assistant Secretaries to act as Secretary for the Bureau.

Article TWENTY

(a) The Secretary General, Assistant Secretaries and the personnel of the Secretariat shall carry out their duties in full independence and in the common interest of the Organization's member countries, and they are not permitted in the performance of their duties to seek or accept instructions from any governmental or non-governmental body. They are required to refrain from any action inconsistent with their duties and, before assuming their posts, to swear to respect the obligations of their office and not to disclose its secrets during or after their service except in cases when the Organization gives its permission. The members of the Organization must respect the neutrality of the Secretary General, the Assistant Secretaries and all the Secretariat's personnel and refrain from trying to influence any of them in any way whatever.

(b) The Secretary General and the Assistant Secretaries shall enjoy in the territories of the Organization members all diplomatic immunities and privileges, while the remaining personnel of the Organization shall enjoy the immunities and privileges necessary for the performance of their tasks and functions in freedom and independence.

(c) The Council shall fix the salaries, allowances and grants to be paid to the Secretary General and the Assistant Secretaries and likewise the administrative and financial bases on which they are treated.

Fourth: The Judicial Board

Article TWENTY ONE

A Judicial Board shall be attached to the Organization, the manner of whose formation and the bases of whose organization shall be agreed upon by the signatories to this Agreement in a special Protocol attached to the Agreement. Hereinafter, it shall be referred to as the Board.

Article TWENTY TWO

The judges of the Board shall be chosen from persons whose impartiality is not in doubt and who fulfill the necessary conditions for holding the highest judicial positions in their countries, or are jurists of international repute. The Council shall appoint the judges in accordance with the conditions and procedures laid down in the Protocol of the Board.

The judges shall take an oath of impartiality, integrity and respect for the Protocol of the Board before the Council.

Article TWENTY THREE

(1) The Board shall be competent to consider the following disputes:

    (a) Disputes relating to the interpretation and application of this Agreement and the implementation of the obligations arising from it.

    (b) Disputes which arise between two or more members of the Organization in the field of petroleum operations.

    (c) Disputes which the Council decides that the Board is competent to consider.

(2) Subject to the approval of the parties to the dispute, the following disputes may be referred to the Board for decision:

    (a) Disputes arising between any member and a petroleum company operating in the territory of the said member.

    (b) Disputes arising between any member and a petroleum company belonging to any other member.

    (c) Disputes arising between two or more members of the Organization, other than what is provided for in paragraph (1) of this Article.

Article TWENTY FOUR

The judgements of the Board shall be considered final, binding and *res judicata* on the parties to the dispute and shall be enforceable *per se* in the territories of the members.

The party concerned shall present the judgement to the local authority responsible for execution, and the competent local authorities shall, after verifying that the document forwarded is official, execute the judgement.

Article TWENTY FIVE

The judges of the Board shall enjoy all diplomatic immunities and privileges in the territories of the Organization's members.

D. Budget of the Organization

Article TWENTY SIX

The Secretary General shall draw up the draft of the Organization's annual budget and submit it to the Council via the Bureau not later than the end of September of the year preceding the implementation of the budget. If the Council has not approved the budget by the beginning of the financial year, the budget of the previous year shall be effective on a monthly basis until the Council approves the new budget.

Article TWENTY SEVEN

The members of the Organization shall contribute to the budget of the Organization in equal shares.

E. General Provisions

Article TWENTY EIGHT

The members of the Organization shall cooperate closely with its organs, coordinate their petroleum policies to the extent necessary to achieve its aims, take all necessary measures to assist the Organization in performing its tasks and to meet their obligations arising out of this Agreement, and refrain from any action which might impede the execution of the Organization's affairs and projects.

Article TWENTY NINE

The members' policies relating to petroleum affairs shall be considered to have a mutual importance, and on this basis, the members undertake to consult among themselves and within the framework of the Organization so as to coordinate their attitudes and such measures as they may take in respect of current situations and circumstances in the petroleum industry.

Article THIRTY

If one of the members is faced with a sudden and serious emergency, the member concerned is entitled to take the necessary precautionary measures, taking all possible care to ensure that such measures do not exceed the minimum required to confront the difficulties that have arisen and do not affect the continued functioning of the Organization's activities.

The member concerned shall immediately inform the Chairman of the Council of the precautionary measures which it has taken, and, should the situation require, the Chairman shall convene the Council for an extraordinary meeting to consider the matter.

Article THIRTY ONE

The organs of the Organization shall endeavor not to do harm to the internal stability of the Organization's members and shall refrain from any action liable to disturb their foreign relations.

Article THIRTY TWO

This Agreement does not oblige any member to disclose information whose disclosure would threaten its own essential security interests.

The members undertake to confine the withholding of information under the preceding paragraph within the narrowest possible limits.

Article THIRTY THREE

In the application of the provisions of this Agreement, a year shall mean the solar year running from 1 January to 31 December inclusive.

Article THIRTY FOUR

The Organization shall, by way of the Secretariat, enter into appropriate contacts with the organs of the United Nations and its specialized agencies, as well as with other organizations.

Article THIRTY FIVE

The Protocols which the parties to this Agreement add to this Agreement shall be considered an integral part of it.

Article THIRTY SIX

This Agreement shall remain in effect for an indefinite period and the amendment of its provisions may be considered every ten years or upon the request of half the members. Such amendment shall require a decision of the Council carried by a majority of three-quarters of the votes of all the members.

Article THIRTY SEVEN

(a) This Agreement shall be ratified by the signatory parties in accordance with their respective constitutional procedures, and the orginal instruments of ratification shall be deposited with the Government of the State of Kuwait within one month of the signing of this Agreement. Copies of them shall be kept by the members.

(b) This Agreement shall come into effect on the first day of the month following the date on which all members have deposited the instruments of ratification; however, if the deposition is completed in the second half of the month, this Agreement shall come into effect as of the beginning of the second month following.

For the	For the	For the
GOVERNMENT OF	GOVERNMENT OF	GOVERNMENT OF
THE KINGDOM OF	THE STATE OF	THE KINGDOM OF
SAUDI ARABIA	KUWAIT	LIBYA

Appendix 2

Empirical Evidence of Arab Community

The methods and results of some empirical investigations into structural Arab community are reported below. Several approaches based on different assumptions about the relations between Arab countries guided the choice of analytical techniques. The null hypothesis for the first technique, comparison of means tests for a number of individual attribute and behavioral variables, is that there is no significant difference between the OAPEC countries considered as a group and a group composed of an equal number of non-OAPEC Arab countries.

OAPEC Countries

Algeria
Bahrain
Egypt
Iraq
Kuwait
Libya
Qatar
Saudi Arabia
Syria
United Arab Emirates (UAE)

Non-OAPEC Arab Countries

Jordan
Lebanon
Mauritania
Morocco
Oman
Somalia
Sudan
Tunesia
North Yemen (ARY)
South Yemen (PDRY)

The non-OAPEC group is made up of Arab League members, in order of their admission to the league, up to ten members (excluding Djibouti), so that the groups would be of equal size. Variables relating to oil production and oil reserves were not used because these are the characteristics that define the OAPEC subset of the Arab League. Next, I assumed that other similarities besides oil linked various Arab countries, and I used cluster analysis to discover these groups of similar countries.

Finally, simple correlations of two economic variables, the consumer price index and an index of money supply changes, were done for each pair of countries with each variable, over a recent time period. Here I assumed that, if economic interdependence did exist among some Arab countries, it would be evident through similar movements in these economic indicators for the interdependent pairs.

Nine different attribute and behavioral characteristics were used as the independent variables in a set of difference of means tests reported in Table 28. The first four variables are population attributes: ethnic and religious composition and, as a measure of economic development, the percentage of each country's work force engaged in agricultural occupations. Arab countries with few non-Muslims and populations nearly or actually monosectarian should be internally more stable than countries such as Lebanon which have powerful religious minorities. Extension of this thesis to the international level would predict that more Arab "community" exists among nations with high proportions of Muslims in their populations, and that community would be even greater if those Muslims were of the Sunni sect. The minimum and maximum values of the percentage of Sunni Muslims in the populations of the members of the two groups of Arab nations are shown below.

	Minimum %	Maximum %
OAPEC	40	99
non-OAPEC	25	99

If one accepts Hourani's definition of "Arab" as ethnic Arab Sunni Muslim, the OAPEC group conforms most closely to the modal Arab type even in the least favorable case. In the matter of ethnic and religious composition, the populations of the OAPEC member states approach the norm of ethnic Arab Muslims more closely than do the populations of the non-OAPEC states. The mean proportion of Sunni Muslims is not significantly different for the two groups, but the least favorable OAPEC case has a higher proportion than the corresponding non-OAPEC nation.

Interestingly, despite the presence of highly developed mineral extraction industries in the OAPEC countries, the two groups are not significantly different with regard to the mean proportions of their respective working populations engaged in agriculture. The oil industry is one of the most capital-intensive; the multinational oil corporations have always preferred to employ home country workers in technical and managerial positions. The home country tendency to monopolize managerial control and its apparent lack of faith in native technical competence have been as much the foundation for host country complaints of exploitation as the low return formerly received for their depleting oil and gas resources. Indeed, the presence or absence of

TABLE 28 Results of Difference of Means Tests

	Variable	Mean	Standard Deviation	F	p
OAPEC	% Sunni	78.50	20.20	1.97	.3253
non-OAPEC		80.80	28.50		
OAPEC	% non-Muslims*	4.30	4.11	14.05	.0005
non-OAPEC		9.60	15.41		
OAPEC	% Arabic speaking*	94.80	7.90	23.82	.0001
non-OAPEC		66.90	38.55		
OAPEC	% farmers	34.00	20.17	1.43	.6022
non-OAPEC		67.50	24.13		
OAPEC	% arms from USA*	6.60	13.96	5.04	.0245
non-OAPEC		21.80	31.33		

Table 28 *continued*

	Variable	Mean	Standard Deviation	F	p
OAPEC	% arms from Eastern European	41.20	47.49	1.51	.5496
non-OAPEC	bloc	32.40	38.66		
OAPEC	arms dependence on a bloc power	- 0.38	0.62	1.20	.8085
non-OAPEC		- 0.11	0.56		
OAPEC	Arab trade preference	4110.70	1486.00	2.02	.3102
non-OAPEC	index	4999.50	2111.20		
OAPEC	political- economic	7.90	3.96	1.39	.6795
non-OAPEC	index	7.88			

Sources: For Sunni, non-Muslim, Arabic-speaking, and farmer percentages: Central Intelligence Agency, *National Basic Intelligence Factbook* (Washington, D.C., July 1977); *The Middle East Yearbook, 1977,* Centre for Middle Eastern and Islamic Studies, University of Durham (England, 1976). For percentages of arms from the U.S. and of arms from the Soviet Union: United States Arms Control and Disarmament Agency, "World Military Expenditures and Arms Transfers 1965-1974" (Washington, D.C., 1975), pp. 74-75; arms dependence is an index taking the difference between the two previous variables. For Arab trade preference index: *Arab Business Yearbook,* "Customs and Tariffs" (London, 1976). The index ran from 1111 to 7777, with the low end reflecting preferences to Arab partners and restrictions on trade with outsiders, and the high end reflecting preferences to outsiders and trade restrictions on Arab partners. The value 4444 was assigned to countries that gave no special preference to any partner. For the political economic index, each country was evaluated and assigned points according to its position on three scales: the number of political parties (1 = 1, 2 or more = 2, 0 = 4), governmental or private control of banking facilities (government = 1, private = 4) and the type of regime (military council = 1, elected president = 2, and monarchy = 4). Information for scoring each was obtained using *The Middle East and North Africa* (London, 1976 and 1977).

*Denotes variable on which a significant difference exists between the two groups.

an oil industry does not appear to have stimulated occupational diversification in the Arab oil-exporting countries. Approximating a definition of development using as an index the percentage of the population engaged in agriculture, all the Arab countries are "developing." Islands of advanced industrial and service capacity are distributed to about the same extent in both the oil-exporting and the non-oil-exporting groups.

Of the three variables which deal with arms dependence, only one indicates a significant difference between the two subgroups of the Arab League. OAPEC member countries have historically spent less of their military budgets on armaments from the United States than the other countries in the Arab League.[1] Both groups of countries have bought arms from the Soviet Union; Western Europe supplied one-fifth or more of the dollar value of armaments to fourteen of the twenty Arab League nations,[2] and over half to eight of these.[3] Most Arab nations have diversified their arms sources and are continuing to do so. There were few cases of dependence on a single source for armaments before 1975. Egypt was a notable exception. And diversification of arms sources continues.

Variable eight is a trade preference index that runs from 1111 to 7777, with 3333 indicating free trade. Below the free trade value are countries that have selective preferences for some or all Arab countries, while values above free trade score indicate restrictions on trade with other Arab countries. The highest score, 7777, shows that the country restricts trade with Arab countries while extending preferences to one or more outsiders. The most common anti-Arab tariff policy is followed by countries extending preferences to the EEC, possibly as a bid for eventual inclusion. Two of these countires, Mauritania and Tunisia, have close ties to France. A single country, Saudi Arabia, charges preferential rates only to Arab countries. The modal score on the index was 4444 (twelve countries) showing uniform tariffs, usually low, on all entering goods regardless of source. For the seven past and/or present members of the Arab Economic Union (Arab Common Market), the mean score was 3809, only a little below the modal value and not significantly different from either test group mean.[4] Thus OAPEC, the non-OAPEC group, and the Arab Common Market group show little difference in their trade policies toward fellow Arab states.

The final variable tested in this set of analyses was an index of political and economic structure. Both test groups of Arab countries were similarly diverse in their varieties of political and economic regime types. Neither the OAPEC group nor the other members of the Arab League viewed as a group has an advantage over the other in having members with more compatible political and economic structures.

The next series of analyses used a clustering procedure to group the entire set of Arab League members according to their similarities on sets of related variables.[5] Those variables marked with an asterisk were examined individually in the previous set of analyses. Four sets of variables were used to partition the member countries of the Arab League:

1. A religious-ethnic set consisting of three variables:
   a. The percentage of Sunni Muslims in the populations*
   b. The percentage of non-Muslims in the populations*
   c. The percentage of Arabic-speaking in the population*

2. An economic development set consisting of four variables.
   a. Gross national product (GNP)
   b. GNP per capita
   c. Population
   d. Percentage of the working population in argiculture*

3. A trade dependence set consisting of four variables:
   a. Arms dependence on a core power*
   b. The Arab trade preference index*
   c. An import dependence index for 1970-73 and
   d. An export dependence index for the same period. These years were chosen to minimize missing values on as many cases as possible, as cases with missing data are omitted by the clustering program.[6]

4. A militarization set consisting of two variables:
    a. Arms dependence on a core power*
    b. Annual military expenditures; two sets of clusters were performed, one using military expenditures for 1974 alone (to minimize missing values) and the other using data for the four-year period 1970-1973.

The clustering procedure is based on a linear or first-order model that treats the value of each variable as a measure of distance on a continuum.[7] The procedure first assigns each case (country) to its own cluster. Then it begins to group the cases by combining the two one-member clusters that are closest together on the distance continuum. Next it combines the two of the remaining groups, including the two-case group produced by the previous step, that are the closest together, and so on. The analyst may choose to stop the procedure after a certain maximum distance has been reached or to continue until only one group consisting of all the cases is formed.[8]

A single variable clustering procedure can be done quite easily "by hand" and provides a simple one-dimensional picture of the way the cases relate to one another over the range of the variable. More interesting results come about when groups of related variables are used simultaneously in a clustering procedure.[9] This produces subgroups of the set of cases corresponding most closely to a multidimensional array of the cases with regard to these related attributes and behaviors. The multivariate clustering, despite its instability, reflects actual characteristics of ecological communities. Indeed, instability in clustering outcomes often corresponds to genuine fluctuations in the degree of attribute and behaviorally dependent structural community among a group of countries.[10]

Table 29 shows the number and composition of clusters formed using the first set of variables, those dealing with the ethnic and religious compositions of the populations of Arab League countries. In these analyses, cluster formation is carried out as long as the standardized distance between the observations farthest apart in any single cluster is less than 1.0. The variables in set 1 divide the Arab League nations into five groups. Steps 2-5 in the procedure formed four two-country clusters with distance values of zero in each. Except for the Kuwait-Saudi Arabia set, these two-country sets include one OAPEC country and one Arab League member not belonging to OAPEC: Algeria-Mauritania, Libya-People's Democratic Republic of Yemen and Qatar-Arab Republic of Yemen. The largest cluster contains eight of the ten OAPEC members plus four non-OAPEC members of the Arab League. The cluster least distant from this one is the three-member cluster that includes the other two OAPEC members and another Arabian Gulf state. This set of clusters confirms the result obtained previously, that is, that OAPEC members resemble one another more closely than do the non-OAPEC members of the Arab League with regard to the religious and ethnic characteristics of their populations. For example, Algeria, similar in its colonial heritage and geographical position as well as in its ethnic composition to Morocco and Tunisia, is included in the large cluster with seven other members of OAPEC. Morocco and Tunisia share a more distant cluster with Somalia. This is because of religious similarities between Algeria and the other members of cluster 1 and because the ethnic composition of Algeria is closer to that of the others in this cluster than is the ethnic composition of either Morocco or Tunisia.

TABLE 29 Clusters Based on Religious and Ethnic Composition

1	2	3	4	5
Algeria*	Bahrain*	Lebanon	Morocco	Sudan
Mauritania	Iraq*		Tunesia	
Egypt*	Oman		Somalia	
Jordan				
Libya*				
ARY				
Qatar*				
PDRY				
Kuwait*				
Saudi Arabia*				
UAE*				
Syria*				
D = 0.73	D = 0.43	D = 0	D = 0.42	D = 0

D = greatest within-cluster distance in standardized units.
* = OAPEC member.

The set of indicators of economic development (see Table 30) breaks the Arab League countries into five groups that do not coincide very well with the religious-ethnic clusters. In this set of clusters, group 5 contains small, poor nations that are generally less developed economically than the rest. One OAPEC member, Syria, is in this cluster. Group 4 contains small, rich countries. All of them are members of OAPEC. Saudia Arabia and Egypt each form a single cluster. Saudi Arabia has a very large gross national product due to its preeminance as an oil exporter. Egypt's population is very much greater than that of any other Arab country, and it is

TABLE 30 Clusters Based on Indicators of Economic Development

1	2	3	4	5
Jordan	Egypt*	Saudi	Bahrain*	Mauritania
Lebanon		Arabia*	Qatar*	Syria*
Oman			UAE*	Somalia
Tunesia			Kuwait*	ARY
Algeria*				PDRY
Iraq*				Sudan
Morocco				
Libya*				
D = 0.58	D = 0	D = 0	D = 0.35	D = 0

D = greatest within-cluster distance in standardized units.
* = OAPEC member.

diversified economically as well. The first cluster is actually a "middle" group, one whose members are not so poor as the countries in group 5 nor so rich as those clusters 3 and 4. Those countries in the first cluster that are not oil exporters have diversified economies and some industry such as mining (Jordan) or banking (Lebanon). The oil exporters in cluster 1 either have small populations and little economic diversification (Libya) or relatively large populations existing at low stand- ards of living compared to the populations of other OAPEC countries. These nations, Iraq and Algeria, are among the most capital hungry of the oil-exporting group.

Again, OAPEC members stand out against most of the rest of the Arab League for one or more reasons relating to economic development: they are richer or larger and more highly developed economically than the others. The OAPEC members belong- ing to the large "middle class" of group 1 are either underdeveloped economically or do not have national incomes sufficiently high to support both welfare and develop- ment needs adequately.

The third set of indicators used in this series of multivariate cluster analyses is the one measuring trade dependence. Due to missing data for Bahrain, the United Arab Emirates, Qatar, and the two Yemens, only fifteen countries are grouped. Table 31 shows the distribution of the fifteen nations across six clusters, three of which have a single member. From this one can see that the Arab League countries are extremely diverse in their trading patterns. All the members of cluster 1, 2, and 5 have a score of 4444 on the Arab trade preference index, which means they have a uniform tariff structure. Cluster 1, Algeria, was also highly dependent on a core power for its arms; and it sent half its exports to only two countries. A similar proportion of its imports orginiated in only two countries. Algeria was a highly trade dependent country across the board.

TABLE 31 *Clusters Based on Trade Dependence*

1	2	3	4	5	6
Algeria*	Libya* Morocco Oman	Mauritania Tunesia	Somalia	Egypt* Kuwait* Lebanon Syria*	Iraq* Jordan Sudan Saudi Arabia*
D = 0	D = 0.55	D = 0.45	D = 0	D = 0.82	D = 0.86

D = greatest within-cluster in standardized units.

* = OAPEC member.

The countries in cluster 2 have similar characteristics except that their import and export trade dependence, although still substantial, is somewhat less than that of Algeria, between 40 percent and 50 percent. Cluster 2 countries differ from Algeria in that their dependence on a core power for military supplies was low, even for Libya, whose score on the index was slightly below 30 percent.

Mauritania and Tunesia make up cluster 2. These countries discriminate against Arab trading partners while favoring non-Arab countries through their tariff structures. Arms dependence was moderate for the two countries, but both imports and exports are concentrated among a few significant partners.

Somalia is the only member of cluster 4. It resembles the members of cluster 3 very closely except that its export dependence is very high; over 70 percent of Somalia's exports went to only two countries.

Cluster 5 countries are arms dependent to varying degrees. Egypt and Syria scored over 95 percent on the index measuring dependence on a core power for armaments. Lebanon was minimally dependent during the period covered by the index, while Kuwait bought no arms from either core power. The four countries in cluster 5 were only moderately dependent with regard to their sources of imports, but export dependence ranged from a low of 27 percent for Lebanon to a high of 43 percent for Egypt, a low-to-moderate range on the index. Egypt has a complex tariff system granting preferences to some Arab countries and to some outsiders as well. The rest of the countries in this cluster have uniform tariffs.

The four members of cluster 6 are the least import and export dependent countries in the Arab League. Each has a tariff structure that favors Arab countries only. However, each one scored 45 percent or more on the arms dependency index, decreasing in one critical area their general dependence on trading partners.

In summary, clustering Arab League countries according to trade policy and trade dependence produced one cluster of relatively trade independent countries that actively seek to increase intra-Arab trade; one cluster of moderately trade dependent countries with effectively neutral policies in regard to intra-Arab trade; and four clusters of highly trade dependent countries whose policies toward intra-Arab trade range from neutrality to discrimination. It should be noted that none of the multi-member clusters is exclusively made up of OAPEC or non-OAPEC countries.

The final set of indicators used in a cluster analysis measured military expenditures and arms dependence on a core power. The clustering procedure was run twice: once using military expenditure data for 1973 only in order to minimize the number of missing cases, and the second time using military expenditure data for the four-year period 1972-76. Table 32 shows the results of the two runs. Allowing for the smaller number of cases in the second analysis, the results were similar in each except for the changed position of Iraq. In each case the level of military spending appears to be the most critical factor allocating cases to clusters. In 1973 Egypt spent more than twice the amount spent by any other Arab country on its military establishment. For the entire period, Saudi Arabia spent the most, Egypt about half that and all the other countries less than half of the amount Egypt spent. Iraq had the third-highest military budget in 1973, but by 1976 its military expenditures had fallen below those of Kuwait.

The results of the four multivariate cluster analyses show that there is no distinct OAPEC cluster on any of these sets of variables. But, in spite of this, members of OAPEC are distinguishable among the Arab nations for reasons other than their oil resources. First, in spite of variations among them, the members of OAPEC reflect a high degree of religious and ethnic similarity. This is shown by their grouping exclusively in the first two clusters in Table 29. This can be important in issue areas where religious or ethnic factors are of some significance, such as the Arab position on Israel. Egyptian leadership in the attempt to come to some accommodation with the state of Israel cannot be discounted as the action of a peripheral Arab state, even though Egypt is widely regarded as a traitor to the Arab cause by many other Arab states.

Ethnically, religiously, and geographically as well, Egypt is one of the core nations of the Arab League. Its position in this regard makes it very hard to ignore, even when Egyptian policy is at variance with Arab policy.

TABLE 32 *Clusters Based on Militarization*

Cluster Run 1			Cluster Run 2		
1	*2*	*3*	*1*	*2*	*3*
Algeria*	Egypt	Saudi Arabia*	Algeria*	Egypt*	Saudi Arabia*
Libya*		Iraq*	Libya*		
Morocco			Morocco		
Jordan			Kuwait*		
Lebanon			Jordan		
Sudan			Lebanon		
Oman			Sudan		
Qatar*			Tunesia		
Tunesia			Oman		
PDRY			Syria*		
ARY			Iraq*		
UAE					
Kuwait*					
Syria*					
D = 0.14	D = 0	D = 0.65	D = 0.85	D = 0	D = 0

* = OAPEC member.
D = the greatest within-cluster distance in standardized units.

If the conformity of one's national population to the modal Arab type increases a nation's policy influence, being in a strong economic position as well can only help. The members of OAPEC have the ability to be policy implementers as well as agenda setters in the region. In addition, the six OAPEC members making up clusters 2-4 in Table 30 are the chief Arab exporters of capital or labor to other Arab nations. Egypt's contribution is mainly labor; the five Arabian Gulf nations are capital exporters. The position of these OAPEC members as Arab "core nations" is strengthened because of their roles in the proliferation of transnational exchanges that create channels for interdependence in the region.

Strangely enough, OAPEC members are among the least trade dependent of the Arab nations, in spite of their commodity dependence on petroleum. But petroleum is an almost unique commodity in international trade.[11] Once extraction capacity has been installed, the flow can be regulated at any level up to capacity consonant with national policy without creating widespread unemployment in the commodity sector.[12] Storage of crude petroleum is costless to the producer, who simply leaves it in the ground.[13] And instead of spoiling and becoming worthless, petroleum in the ground promises to increase in value over time.[14] Thus petroleum exporters need regard no other internal demand than their need for income in determining production levels up to extraction capacity. Too, even though petroleum and natural gas form the major exports of eight of the ten members of OAPEC[15] the world-wide distribution of demand makes export partner dependence unnecessary. For these reasons the majority of those OAPEC countries for which data was available fall in the clusters of nations exhibiting the least trade dependence (see Table 31).

A multiple regression analysis was done to test the effects of import and export concentration on the Arab trade preference index. No significant relationship was found.

With regard to militarization, only oil-exporting Arab nations have sufficient independent resources to support levels of militarization of regional importance. However, military expenditures in the Arab countries have continued to increase. Countries such as Kuwait, whose military spending was close to zero ten years ago, are spending more and more. Arms sources continue to diversify because the Arab countries are reluctant to become dependent on either core power. An informed source told me, in July 1980, that Saudi Arabia had resumed secret negotiations with the Soviet Union to obtain arms as a way to lessen its military dependence on the United States, although later press reports have indicated that these talks did not go very far. Unfortunately, military spending is increasing in many places in the region, contributing to local instability as well as to the likelihood of the Middle East becoming the battleground for a proxy war in the 1980s.

The third set of analyses done here employed a methodology that promised to go beyond the first level of structural community to capture its second level effects. These effects have been equated with international interdependence and are in part the results of transactions which form the first level of structural community. The methodology was first used by Richard Rosecrance and his colleagues to test theories of variations in international interdependence over time.[16] Rosecrance et al. constructed their research design to capture the sensitivity component of interdependence.[17] They distinguished between two types of interdependence created by different methods of measurement. When interdependence is measured by transactions, "the flow of money, men [and] goods . . . " it is horizontal interdependence. Vertical interdependence is measured by the responses "of one economy to another in terms of changes in factor prices."[18] It should be noted that only vertical interdependence is interdependence in the sense meant by scholars such as Richard Cooper, Robert Keohane, and Joseph Nye.[19] They regard transactions simply as pathways along which interdependence might develop rather than as interdependence itself. And, in spite of calling measures of transactions horizontal interdependence, Rosecrance et al. also say that "horizontal interdependence . . . implies only connectedness."[20]

Having established the concept of vertical interdependence as sensitivity to movements in factor prices, the Rosecrance group suggests two "obvious" measures for it: first, a "correlation of the movement of factor-price indices between . . . two societies; and, second, a correlation of the changes in such movements."[21] The methodology was tested using the consumer price index (CPI), an index of wages and prices and an index of manufacturing. Each index was correlated individually between pairs of American cities for various time periods. The values on the various indices were generally highly intercorrelated, although not to the same extent for each index or for each time period considered. Because of the positive results for urban societies known to be integrated through a national government, Rosecrance et al. concluded that their methodology was capable of capturing the results of interdependence between different nations. They then used the correlation procedure on similar data from six developed countries over several time periods to test four theories concerning patterns of interdependence across time. Again, the correlation procedure appeared to indicate that high levels of interdependence had existed between pairs of developed countries, although the precise levels of interdependence had varied over time.

The work reported here employs the CPI, an index used by Rosecrance, and a quasi-index of each nation's money supply created by the log transformation of raw data values in local currency for the money supply at each time point in each country.[22] Although governmental intervention in interest and foreign exchange rates makes the use of these indicators questionable, every student of the Federal Reserve Board knows that the money supply is much less manipulable. According to an extension of the purchasing power parity doctrine,[23] a nation's money supply, in the absence of effective foreign investment controls, should increase as the money supply of a nation with which it is economically interdependent. Since the growth of banking facilities has been extensive in the Arab region, the money supply indicator seemed a promising way to capture international capital flows.

Tables 33 and 34 report the correlation coefficients between pairs of countries on each of these indicators. The CPIs are highly intercorrelated even where the overall secular trend of rising prices might at first be expected to have less effect: on the relatively isolated economies of the Yemens. But a factor analysis of the correlation matrix produced only one factor, reflecting the monolithic pattern of inflation throughout the area.[24] The results reported for the correlation of the money supply index do not vary in such a unidirectional way, but, even so, the correlations produced are very high on the whole. Factor analysis of this correlation matrix gave two factors which are shown in Table 35 in both their unrotated and rotated forms. The rotated factor patterns show most of the oil exporters in a single large group that also includes some recipients of their foreign assistance. The second factor can be difined as Somalia. Lebanon and the United Arab Emirates are not closely associated with either factor.

Even if very high standards are chosen as cut points in the determination of which countries are interdependent and which are not, the results for both analyses are dismayingly high. According to these results few pairs of Arab countries are not economically interdependent. If Somalia were omitted from the analysis, nearly every possible pair of Arab countries could be said to be interdependent. It might be the case that these countries really are closely interdependent. Trade, foreign aid, arms, and people pass regularly from one Arab country to another.[25] Yet one might reasonably expect to see a greater variety of relationships than "very interdependent" and "very, very interdependent." Accordingly, I retested the methodology using a test group and a control group of countries and discovered that it is not entirely reliable. False positive results were obtained with great frequency; it also appeared that gradations in the amount of international sensitivity between any pair of countries is probably not definable using the methodology.[26] The best that can be said is that the methodology provides an initial screening of possible pairs of countries for indications that they are interdependent. The extremely positive results reported here are not conclusive. However, had the results shown little correspondence in the movement of the indicators tested over time between pairs of Arab countries, we would have been able

TABLE 33 Pearson's r for Consumer Price Index

	AL	BA	EG	IQ	JO	KU	LE	LI	MA	MO	SA	SO	SU	SY	TU	YS	YA
AL	1.000	.978	.984	.965	.967	.958	.969	.942	.989	.966	.984	.961	.907	.977	.975	.899	.972
BA	.978	1.000	.992	.975	.984	.989	.976	.939	.986	.993	.971	.972	.966	.992	.971	.966	.986
EG	.984	.992	1.000	.976	.983	.977	.969	.932	.992	.987	.983	.976	.948	.987	.976	.949	.973
IQ	.965	.975	.976	1.000	.987	.938	.946	.917	.964	.969	.964	.944	.929	.974	.975	.941	.936
JO	.967	.984	.983	.987	1.000	.971	.976	.928	.960	.984	.957	.944	.935	.981	.971	.959	.946
KU	.958	.989	.977	.938	.971	1.000	.989	.905	.962	.992	.945	.970	.958	.985	.950	.977	.992
LE	.969	.976	.969	.946	.976	.989	1.000	.823	.973	.990	.917	.892	.962	.976	.941	.981	.986
LI	.942	.939	.932	.917	.928	.905	.823	1.000	.931	.914	.912	.896	.928	.948	.934	.925	.931
MA	.989	.986	.992	.964	.960	.962	.973	.931	1.000	.975	.984	.974	.926	.976	.940	.913	.972
MO	.966	.993	.987	.969	.984	.992	.990	.914	.975	1.000	.955	.983	.965	.984	.962	.973	.986
SA	.984	.971	.983	.964	.957	.945	.917	.912	.984	.955	1.000	.983	.912	.965	.958	.893	.957
SO	.961	.972	.976	.944	.944	.970	.892	.896	.974	.964	.983	1.000	.934	.959	.940	.910	.967
SU	.907	.966	.948	.929	.935	.958	.962	.928	.926	.965	.912	.934	1.000	.967	.945	.975	.938
SY	.977	.992	.987	.974	.981	.985	.976	.948	.976	.984	.965	.959	.967	1.000	.983	.971	.980
TU	.975	.971	.976	.975	.971	.950	.941	.934	.940	.962	.958	.940	.945	.983	1.000	.944	.966
YS	.899	.966	.949	.941	.959	.977	.981	.925	.913	.973	.893	.910	.975	.971	.944	1.000	.960
YA	.972	.986	.973	.936	.946	.992	.986	.931	.972	.986	.957	.967	.938	.980	.966	.960	1.000

Source: International Monetary Fund, *International Financial Statistics*, various volumes. Revised figures were used when available.
Note: See notes for table 34, p. 197.

TABLE 34 *Pearson's r for Ln (Money Supply)*

	AL	BA	EG	IQ	JO	KU	LE	LI	MA	MO	QT	SA	SO	SU	SY	TU	UA	YS	YA
AL	1.000	.941	.988	.974	.971	.977	.986	.969	.942	.989	.984	.974	.561	.983	.979	.981	.955	.967	.979
BA	.941	1.000	.931	.893	.920	.946	.858	.922	.862	.923	.956	.928	.491	.906	.911	.900	.976	.959	.946
EG	.988	.931	1.000	.991	.991	.980	.978	.967	.957	.996	.990	.988	.526	.989	.993	.979	.988	.979	.966
IQ	.974	.893	.991	1.000	.988	.959	.931	.945	.956	.988	.978	.981	.531	.983	.989	.968	.977	.969	.941
JO	.971	.920	.991	.988	1.000	.973	.978	.945	.941	.980	.990	.994	.533	.968	.978	.951	.970	.983	.977
KU	.977	.946	.980	.973	.973	1.000	.967	.953	.926	.972	.991	.979	.557	.959	.960	.951	.977	.973	.987
LE	.986	.858	.978	.931	.978	.967	1.000	.953	.751	.965	.981	.970	.975	.948	.951	.940	.979	.953	.982
LI	.969	.922	.967	.945	.945	.953	.953	1.000	.949	.980	.951	.926	.588	.978	.973	.985	.963	.932	.927
MA	.942	.862	.957	.956	.941	.926	.751	.949	1.000	.963	.334	.927	.443	.970	.969	.965	.727	.916	.840
MO	.989	.923	.996	.988	.980	.972	.965	.980	.963	1.000	.983	.976	.556	.994	.994	.990	.983	.967	.951
QT	.984	.956	.990	.978	.990	.991	.981	.951	.334	.983	1.000	.994	.552	.970	.972	.960	.991	.987	.992
SA	.974	.928	.988	.981	.994	.979	.970	.926	.927	.976	.994	1.000	.505	.961	.969	.948	.993	.987	.988
SO	.561	.491	.526	.531	.533	.557	.975	.588	.443	.556	.552	.505	1.000	.548	.514	.528	.950	.531	.257
SU	.983	.906	.989	.983	.968	.959	.948	.978	.970	.994	.970	.961	.548	1.000	.993	.989	.958	.957	.928
SY	.979	.911	.993	.989	.978	.960	.951	.973	.969	.994	.972	.969	.514	.993	1.000	.983	.972	.961	.925
TU	.981	.900	.979	.968	.951	.951	.940	.985	.965	.990	.960	.948	.528	.989	.983	1.000	.929	.933	.916
UA	.955	.976	.988	.977	.970	.977	.979	.963	.727	.983	.991	.993	.950	.958	.972	.929	1.000	.980	.989
YS	.967	.959	.979	.969	.983	.973	.953	.932	.916	.967	.987	.987	.531	.957	.961	.933	.980	1.000	.989
YA	.979	.946	.966	.941	.977	.987	.982	.927	.840	.951	.992	.988	.257	.928	.925	.916	.989	.989	1.000

Source: International Monetary Fund, *International Financial Statistics*, various volumes. Revised figures were used when available.

Notes: Correlation coefficients for the consumer price index, yearly, 1969-72, and quarterly, 1973-76. This index was corrected to 1963 = 100.00 for all cases. Correlation coefficients for the money supply, yearly, 1969-72, and quarterly, 1973-76, were obtained using the natural log transformation of each raw data value. The abbreviations in Tables 33 and 34 refer to the following countries:

AL	Algeria
BA	Bahrain
EG	Egypt
IQ	Iraq
JO	Jordan
KU	Kuwait
LE	Lebanon
LI	Libya
MA	Mauritania
MO	Morocco
QT	Qatar
SA	Saudi Arabia
SO	Somalia
SU	Sudan
SY	Syria
TU	Tunesia
UA	United Arab Emirates
YS	Arab Republic of Yemen, capital Sa'ana
TA	People's Democratic Republic of Yemen, capital Aden

TABLE 35 *Loadings on Money Supply Factors*

Country	Rotated Factors		Unrotated Factors	
	1	2	1	2
Algeria	.915	.387	.992	-.039
Bahrain	.884	.334	.942	-.074
Egypt	.930	.337	.998	-.064
Iraq	.919	.359	.984	-.066
Jordan	.919	.359	.989	-.057
Kuwait	.907	.388	.986	-.035
Lebanon	.739	.733	.981	.349
Libya	.882	.414	.975	-.001
Mauritania	.939	.187	.930	-.228
Morocco	.919	.385	.995	-.042
Qatar	.918	.388	.996	-.040
Saudi Arabia	.925	.354	.988	-.073
Somalia	.178	1.027	.598	.853
Sudan	.915	.372	.986	-.052
Syria	.925	.354	.987	-.074
Tunisia	.913	.353	.976	-.069
United Arab Emirates	.764	.710	.993	.318
Arab Republic of Yemen	.914	.366	.983	-.057
People's Democratic Republic of Yemen	.973	.198	.965	-.235

to conclude that they are not interdependent. A supporting analysis showing relatively high levels of economically relevant transactions would increase the likelihood that the correlation results are reflecting economic reality in the region. This would be particularly encouraging if a case like Somalia could be shown to have participated in significantly fewer transactions than the pairs of countries for which the correlations were extremely high.

Notes

1. The data for this variable consists of the percentages of the cumulative expenditures for arms from 1965 to 1974 that were purchased from the United States.

2. They were: Jordan, Kuwait, Lebanon, Libya, Mauritania, Morocco, Oman, Qatar, Saudi Arabia, Sudan, Tunisia, the United Arab Emirates, and both Yemens.

3. The eight were: Kuwait, Lebanon, Libya, Mauritania, Oman, Qatar, Saudi Arabia, and the United Arab Emirates.

4. For heuristic purposes a one-way analysis of variance was done on the three groups. With 2 and 26 degrees of freedom the value for F was 2.52. Critical F for those degrees of freedom is 4.32 at $p = .05$.

5. Data for the indicators was obtained from the following sources: Central Intelligence Agency, *National Basic Intelligence Factbook* (Washington, D. C., July 1977);

The Middle East Yearbook, 1977 (Durham, England, 1976); United Nations, *1976 Statistical Yearbook* (United Nations, 1977); The Institute for Strategic Studies, *The Military Balance* (London, 1968-1977); *Arab Business Yearbook* (London, 1976); United States Arms Control and Disarmament Agency, *Military Expenditures 1965-1974* (Washington, D. C., 1975). Whenever possible the same source was used for each indicator for all cases.

6. The program used was version "N" of the Statistical Package for the Social Sciences obtained from the SAS Institute, P.O. Box 10066, Raleigh, N.C. 27605. The procedure was "Cluster." No case weights were used. Distances were standardized as follows:

$d = n^{-1} sum_i [d(x_i, \bar{x})]$, where i = ith observation vector and n = the number of observations in the data set.

7. See J. A. Hartigan, *Clustering Algorithms* (New York, 1975), Chaps. 1-3.

8. Clusters reported here were chosen on the basis of having no standardized within-group distance larger than 1.0.

9. There is a problem in that using more than one variable in a clustering procedure results in "unstable" clusters, that is, results that change when one variable is removed—or added. Other types of multivariate analysis, such as factor analysis, are similarly unstable. This mathematical instability is a function of the real similarities and differences between cases when different dimensions are used to put them into categories.

10. This can be thought of as issue-based structural community. Cooperation on any issue among several nations with different levels of relevant attributes and which behave differently on relevant indicators will probably vary in that countries that are most similar will share policy preferences.

11. Gold, diamonds, and uranium might be considered similar kinds of internationally traded commodities because of the relative ease in holding them off the market. But there is no real analog to production capacity.

12. Because the petroleum industry is so highly capital-intensive.

13. See Fadhel al-Chalabi, "Pricing of OPEC Crude Oil: A Case for the Valuation of Depletable Resources in Relation to Economic Development," mimeographed (Kuwait, 1977).

14. See Robert Mabro, "Energy Crisis in 1985?" *Middle East Economic Survey (MEES)*, Supplement, 10 April 1978.

15. These are: Algeria, Iraq, Kuwait, Libya, Qatar, Saudi Arabia, Syria, and the UAE. The majority of Bahrain's exports are manufactured products and not petroleum or petroleum products. Egypt does not export much oil compared to its total export package. Syria exports very little oil.

16. Richard Rosecrance, Alan Alexandroff, Wallace Koehler, John Kroll, Shlomit Lacquer, and John Stocker, "Whither Interdependence?" *International Organization* 31 (Summer 1977).

17. Robert O. Keohane and Joseph S. Nye, *Power and Interdependence* (Boston, 1977), Chap. 1.

18. Rosecrance et al., pp. 428-29.

19. Richard Cooper, *The Economics of Interdependence: Economic Policy in the Atlantic Community* (New York, 1968), Chap. 1; and Keohane and Nye, *Power and Interdependence*, Chap. 1.

20. Rosecrance et al., p. 427. The term "interconnectedness" is from Alex Inkeles, "The Emerging Social Structure of the World," *World Politics* 27 (July 1975).

21. Rosecrance et al., p. 428. The article also uses a convergence statistic measuring the degree of equalization over time in the international data, but I do not use that technique in this work.

22. This technique also avoided the problem of standardizing the currencies to one denomination as the log transformations become precentage changes when considered over time. I am indebted to Gordon Smith, professor of economics at Rice University, for his comments on the entire range of indices I considered for this analysis and for his suggestion to log the values for the money supply.

23. Charles Kindleberger, *International Economics*, 5th ed. (Homewood, Ill.; 1973), p. 390.

24. The factor loadings for the consumer price index correlation and factor analysis were:

Algeria	0.060
Bahrain	0.061
Egypt	0.061
Iraq	0.060
Jordan	0.060
Kuwait	0.060
Lebanon	0.060
Libya	0.058
Mauritania	0.060
Morocco	0.061
Saudi Arabia	0.060
Somalia	0.059
Sudan	0.059
Syria	0.061
Tunisia	0.060
ARY	0.059
PDRY	0.060

These values are virtually identical and show no meaningful pattern of variation.

25. See, for example, Ibrahim Sa'ad Eddin, "The Negative Effects of Differences of Income Among Arab Countries on Development in Countries with Low Per Capita Income: The Case of Egypt." Egypt is usually a labor donor. Kuwait accepts foreign labor. More than half the population of Kuwait is of foreign origin, many having immigrated to take employment. North Yemen (ARY) and Jordan are also important labor exporters and Palestinians can be found throughout the region.

26. Mary Ann Tétreault, "Measuring Interdependence," *International Organization* 34 (Summer 1980).

Bibliography

Abu Khadra, Rajai M. "The Spot Oil Market: Genesis, Qualitative Configuration and Perspectives." *OPEC Review 3/4*. (Winter 1979/Spring 1980): 105-15.

Adelman, M. A. *The World Petroleum Market*. Baltimore: Johns Hopkins University Press, 1972.

———. "Is the Oil Shortage Real? Oil Companies as OPEC Tax Collectors." *Foreign Policy* 9 (Winter 1972-3): 69-107.

Ajami, Fouad. "The End of Pan-Arabism." *Foreign Affairs* 57 (Winter 1978/79): 355-73.

Albaharna, Husan M. *The Arabian Gulf States*. 2d rev. ed. Beirut: Librairie du Liban, 1975.

Anabtawi, M. F. *Arab Unity in Terms of Law*. The Hague: Martinus Nijhoff, 1963.

Anthony, John D. *The Middle East: Oil, Politics and Development*. Washington, D.C.: American Enterprise Institute, 1975.

Antonius, George. *The Arab Awakening*. London: Hamish Hamilton, 1938.

Arab Business Yearbook: 1976. London: Graham and Trotman, 1976.

Arab Petroleum Directory,1974-1975. Beirut: Arab Petroleum Directory Association, 1975.

Archer, Jules. *Legacy of the Desert*. Boston: Little, Brown and Company, 1976.

Attiga, Ali A. "The Impact of Energy Transition on the Oil Exporting Countries." Paper presented at the Fourth International Energy Conference, Boulder, Colorado, 17-19 October 1977. Mimoegraphed.

———. "Regional Cooperation in Downstream Investments: The Case of OAPEC." Paper presented at the OPEC Seminar on the Present and Future Role of the National Oil Companies, Vienna, 10-12 October 1977. Mimeographed.

Auda, Abul-Futuh Hamed. *Arab League Systems*. New York: Arab Information Center, 1972.

Axline, William A. "Underdevelopment, Dependency and Integration: The Politics of Regionalism in the Third World." *International Organization* 31 (Winter 1977): 83-105.

Banna, Abdel-Monem el-. *The Arab Economic Unity*. New York: Arab Information Center, 1972.

Barnet, Richard, J. "The World's Resources—Part I." *New Yorker,* 17 March 1980, pp. 45-81.

Becker, A. S.; Hansen, B.; and Kerr, M. *The Economics and Politics of the Middle East*. New York: American Elsevier, 1975.

Bergman, Lars, and Radetzki, Marian. "How Will the Third World Be Affected by OECD Energy Strategies?" *Journal of Energy and Development* 5 (Autumn 1979): 19-31.

Bergsten, C. Fred. "The Threat Is Real." *Foreign Policy* 14 (Spring 1974): 84-90.

Birks, J. S., and Sinclair, C. A. "Egypt: A Frustrated Labor Exporter?" *The Middle East Journal* 33 (Summer 1979): 278-303.

Bill, James A., and Leiden, Carl. *The Middle East: Politics and Power.* Boston: Allyn and Bacon, 1974.

Bill, James, and Stookey, Robert W. *Politics and Petroleum.* Brunswick, Ohio: Kings Court Press, 1975.

Blair, John M. *The Control of Oil.* New York: Pantheon, 1976.

Bloom, John. "Profits? What Profits?" *Texas Monthly,* July 1980.

Bobrow, Davis B., and Kudrle, Robert T. "Theory, Policy and Resource Cartels: The Case of OPEC." *Journal of Conflict Resolution* 20 (March 1976) 3-56.

Brown, Stephen N.; Price, David; and Raichur, Satish. "Public Goods Theory and Bargaining Between Large and Small Countries." *International Studies Quarterly* 20 (September 1976): 393-414.

Cantori, Louis J., and Spiegal, Stephen L. "The Analysis of Regional International Politics: The Integration Versus the Empirical Systems Approach." *International Organization* 27 (Autumn 1973): 465-94.

Carmichael, Joel. *The Shaping of the Arabs: A Study in Ethnic Identity.* New York: Macmillan, 1967.

Caves, Richard E. *International Cartels and Monopolies in International Trade.* Discussion Paper Series. Cambridge, Mass.: Harvard Institute of Economic Research, 1977.

Central Intelligence Agency. *The International Energy Situation: Outlook to 1985.* Washington, D. C., 1977.

Chalabi, Fadhel Al-. "Pricing of OPEC Crude Oil: A Case for the Valuation of Depletable Resources in Relation to Economic Development." Courtesy of George G. Tomeh. Mimeographed.

_____. "Past and Present Patterns of the Oil Industry in the Producing Countries." *OPEC Review* 3/4 (Winter 1979/Spring 1980): 7-20.

_____, and Al-Janabi, Adnan. "Optimum Production and Pricing Policies." *Journal of Energy and Development* 4 (Spring 1979): 229-58.

Chamie, Joseph. "Religious Groups in Lebanon: A Descriptive Investigation." *International Journal of Middle East Studies* 2 (April 1980): 175-87.

Choucri, Nazli. *The International Politics of Energy Interdependence.* Lexington, Mass.: Lexington Books, 1976.

Cobb, Roger W., and Elder, Charles. *International Community: A Regional and Global Study.* New York: Holt, Rinehart and Winston, 1970.

Comanor, William S., and Schankerman, Mark A. "Identical Bids and Cartel Behavior." *Bell Journal of Economics* 7 (Spring 1976): 281-86.

Committee of Experts on Restrictive Business Practices. *Export Cartels.* Paris: Organization for Economic Cooperation and Development, 1974.

Cooper, Richard N. "Economic Interdependence and Foreign Policy in the 1970's." *World Politics* 24 (January 1972): 159-81.

_____ . *The Economics of Interdependence: Economic Policy in the Atlantic Community.* New York: McGraw Hill, 1968.

Crawford, David. *Getting Work in OPEC.* London: The Architectural Press, 1975.

Dailami, Mansoor. "Inflation, Dollar Depreciation, and OPEC's Purchasing Power." *Journal of Energy and Development* 4 (Spring 1979): 326-43.

De Rousiers, Paul. *Cartels and Trusts and Their Development.* Geneva: League of Nations Economic and Financial Section, 1927.

Deutsch, Karl W.; Burrell, Sidney A.; Kann, Robert A.; Lee, Maurice, Jr.; Lichterman, Martin; Lindgren, Raymond E.; Loewenheim, Francis L.; and Van Wagenen, Richard W. *Political Community and the North Atlantic Area.* Princeton: Princeton University Press, 1957.

Doran, Charles F. "Conflict, Cohesion and Coalition Formation in OPEC: A Quantitative Assessment." Paper presented at the 1977 Annual Meeting of the American Political Science Association, Washington, D. C., 1-4 September 1977.

―――. "Energy and United States Foreign Policy: Birth of a New Alliance." Paper presented at the Annual Meeting of the Southern Political Science Association, New Orleans, November 1977.

―――. *Myth, Oil and Politics.* New York: The Free Press, 1977.

Duguid, Stephen. "A Biographical Approach to the Study of Social Change in the Middle East: Abdullah Tariki as a New Man." *International Journal of Middle East Studies* 1 (July 1970): 195-220.

Eckbo, Paul Leo. *The Future of World Oil.* Cambridge, Mass.: Ballinger Books, 1975.

Eddin, Ibrahim Saad. "The Negative Effects of Difference of Income among Arab Countries on Development in Countries with Low Per Capita Income: The Case of Egypt." In OAPEC, *Sources and Problems of Arab Development.* Kuwait: OAPEC, 1980.

Engler, Robert. *The Brotherhood of Oil.* Chicago: University of Chicago Press, 1977.

Fathaly, Omar I., and Palmer, Monte. "Opposition to Change in Rural Libya." *International Journal of Middle East Studies* 11 (April 1980): 247-61.

Fesharaki, Fereidun. "Global Petroleum Supplies in the 1980's: Prospects and Problems." *OPEC Review* 4 (Summer 1980): 27-49.

Fisher, E. M., and Bassiouni, M. Cherif. *Storm over the Arab World.* Chicago; Follett Press, 1972.

Fog, Bjarke. "How Are Cartel Prices Determined?" *Journal of Industrial Economics* 5 (November 1956): 16-23.

Fried, E. R., and Schultz, Charles L. *Higher Oil Prices and the World Economy.* Washington, D.C.: The Brookings Institution, 1975.

Friedland, Edward; Seabury, Paul; and Wildavsky, Aaron. "Oil and the Decline of Western Power." *Political Science Quarterly* 90 (Fall 1975): 437-50.

Friedrich, Carl J., ed. *Community.* New York: Liberal Arts Press, 1959.

Geddawy, A. Kesmat el-. "Arab Companies Established by OAPEC." In OAPEC, ed., *Petroleum and Arab Economic Cooperation.* Kuwait: Organization of Arab Petroleum Exporting Countries, 1978.

Gereffi, Gary. "Drug Firms and Dependency in Mexico: The Case of the Steroid Hormone Industry." *International Organization* 32 (Winter 1978): 237-86.

Haas, Ernst. "The Study of Regional Integration: Reflections on the Joy and Anguish of Pre-theorizing." *International Organization* 24 (Autumn 1970): 607-46.

―――. "Turbulent Fields and the Theory of Regional Integration." *International Organization* 30 (Spring 1976): 173-212.

―――. *The Uniting of Europe.* Stanford, Calif. Stanford University Press, 1958.

Hammond, Paul Y., and Alexander, Sidney S., eds. *Political Dynamics in the Middle East.* New York: American Elsevier, 1972.

Hansen, Roger D. "Regional Integration: Reflections on a Decade of Theoretical Efforts." *World Politics* 21 (January 1969): 242-65.

Harik, Iliya F. "The Ethnic Revolution and Political Integration in the Middle East." *International Journal of Middle East Studies* 3 (July 1972): 303-23.

Hartshorn, J. E. *Objectives of the Petroleum Exporting Countries.* Nicosia, Cyprus: Middle East Petroleum and Economic Publications, 1978.

Hexner, Erwin. *International Cartels.* London: Sir I. Pitman & Sons, 1946.

Hirst, David. *Oil and Public Opinion in the Middle East.* London: Faber and Faber, 1966.

Hottinger, Arnold. *The Arabs: Their History, Culture and Place in the Modern World.* Berkeley, Calif.: University of California Press, 1963.

Hourani, Albert H. *Minorities in the Arab World.* London: Oxford University Press, 1947.

———. *A Vision of History.* Beirut: Khayats, 1961.

Hubbert, M. King. "Energy Resources." In *Resources and Man,* ed. National Academy of Sciences-National Resource Council. San Francisco: Freeman Press, 1969.

Hudson, Michael C. *Arab Politics: The Search for Legitimacy.* New Haven: Yale University Press, 1977.

Humphreys, R. Stephen. "Islam and Political Values in Saudi Arabia, Egypt and Syria." *The Middle East Journal* 33 (Winter 1979): 1-19.

Inglehart, Ronald. "Public Opinion and Regional Integration." *International Organization* 24 (Autumn 1970); 764-95.

Inkeles, Alex. "The Emerging Social Structure of the World." *World Politics* 27 (July 1975): 467-95.

Iskander, Marwan. *The Arab Oil Question.* Beirut: Middle Eastern Economic Consultants, 1974.

Ismael, Tareq Y. *The Middle East in World Politics.* Syracuse, N.Y.: Syracuse University Press, 1974.

Issawi, Charles. *Oil, the Middle East and the World.* Washington Papers, Vol. 1, No. 4. Beverly Hills: Sage Publications, 1972.

Jabber, Paul. "Conflict and Cooperation in OPEC: Prospects for the Next Decade." *International Organization* 32 (Spring 1978): 377-400.

Jacoby, Neil H. *Multinational Oil.* New York: Macmillan, 1974.

Jaidah, Ali M. "Downstream Operations and the Development of OPEC Member Countries." *Journal of Energy and Development* 4 (Spring 1979): 304-12.

Johany, Ali D. "OPEC and the Price of Oil: Cartelization or Alteration of Property Rights." *Journal of Energy and Development* 5 (Autumn 1979): 73-80.

———. "OPEC Is Not a Cartel: A Property Rights Explanation of the Rise in Crude Oil Prices." PhD. dissertation, University of California, Santa Barbara, 1978.

Kaiser, Karl. "The Interaction of Regional Subsystems." *World Politics* 21 (October; 1968): 84-102.

Kanovsky, E. "Arab Economic Unity." In Joseph S. Nye, ed. *International Regionalism.* Boston: Little Brown, 1968.

Katzenstein, Peter J. "International Interdependence: Some Long-Term Trends and Recent Changes." *International Organization* 29 (Autumn 1975): 1021-34.

Kazziha, W. W. *Revolutionary Transformation in the Arab World.* New York: St. Martin's Press, 1975.

Kegley, Charles W., and Howell, L. "The Dimensionality of Regional Integration: Construct Validation in the South East Asia Context." *International Organization* 29 (Autumn 1975): 997-1020.

Kennedy, Michael. "An Economic Model of the World Oil Market." *Bell Journal of Economics* 5 (Autumn 1974): 540-77.

Keohane, Robert O., and Nye, Joseph S. *Power and Interdependence.* Boston: Little, Brown, 1977.

Keohane, Robert O., and Ooms, Van Doorn. "The Multinational Firm and International Regulations." *International Organization* 29 (Winter 1975): 169-209.

Kiernan, Thomas. *The Arabs: Their History, Aims and Challenge to the Industrialized World.* Boston: Little, Brown, 1975.

Klebanoff, S. *Middle East Oil and United States Foreign Policy.* New York: Praeger, 1974.

Kraft, Joseph. "A Letter from OPEC." *New Yorker* (28 January 1980): 68-77.

Krasner, Stephen D. "Oil Is the Exception." *Foreign Policy* 14 (Spring 1974): 63-83.

Kruger, Robert B. *The United States and International Oil.* New York: Praeger, 1975.

Landes, David. *Bankers and Pashas.* New York: Harper and Row, 1958.

Lattu, Onnie P. Remarks Delivered Before the American Petroleum Institute Southern District Division of Production, San Antonio, 7 March 1968.

League of Nations, Department of Economic Affairs, "International Cartels: A League of Nations Memorandum." Lake Success, N.Y. 1947.

Lenczowski, George. "Arab Bloc Realignments." *Current History* 52 (December 1967): 346-51, 384.

———. "Arab Radicalism: Problems and Prospects." *Current History* 60 (January 1971): 32-37, 52.

Levy, Walter J. "Oil and the Decline of the West." *Foreign Affairs* 58 (Summer 1980): 999-1015.

———. "The Years That the Locust Hath Eaten: Oil Policy, and OPEC Development Prospects." *Foreign Affairs* 57 (Winter 1978/79): 287-305.

Lindberg, Leon N., and Scheingold, Stuart A. *Europe's Would-Be Polity.* Englewood Cliffs, N.J.: Prentice Hall, 1970.

Loehr, William. "Collective Goods and International Cooperation." *International Organization* 27 (Summer 1973): 421-30.

Losman, Donald L. "The Arab Boycott of Israel." *International Journal of Middle East Studies* 3 (April 1972): 99-115.

Lowi, Theodore J. "American Business, Public Policy, Case Studies and Political Power." *World Politics* 16 (July 1964): 677-715.

———. "Four Systems of Policy, Politics and Choice." *Public Administration Review* 32 (July/August 1972): 298-310.

Lyons, Richard D. "Carter Energy Plan Is in Place But So Are Many Boobytraps." *New York Times* (25 May 1980): 20E.

Mabro, Robert. "Energy Crisis in 1985?" *Middle East Economic Survey Supplement,* 10 April 1978.

_____. "The Marker and the Market, the Heavy and the Light." *Middle East Economic Survey Supplement*, 18 September 1978.

_____. "OPEC After the Oil Revolution." *Journal of International Studies—Millennium* (Winter 1975): 191-99.

_____ , and Samir Radwan. *The Industrialization of Egypt, 1939-1973*. Oxford: Clarendon Press, 1976.

MacAvoy, Paul W. *The Economic Effects of Regulation*. Cambridge, Mass.: MIT Press, 1965.

McCormick, James M. "Intergovernmental Organizations and Cooperation Among Nations." *International Studies Quarterly* 24 (March 1980): 73-98.

MacDonald, Robert W. *The League of Arab States*. Princeton: Princeton University Press, 1965.

Mallakh, Ragaei El. "Industrialization in the Arab World: Obstacles and Prospects." In Naiem Sherbiny and Mark Tessler, eds. *Arab Oil: Its Impact on the Arab Countries and Global Implications*. New York: Praeger, 1976.

Mazrui, Ali A. "The Barrel of the Gun and the Barrel of Oil in the North-South Equation." Working Paper Number 5. World Oil Models Project (1978).

Mead, Walter, J. "An Economic Analysis of Crude Oil Price Behavior in the 1970's." *Journal of Energy and Development* 4 (Spring 1979): 212-28.

The Middle East and North Africa. London: Europa Publications, 1976.

The Middle East Yearbook, 1977. Durham, England: The Centre for Middle Eastern and Islamic Studies, 1976.

Mikdashi, Zuhayr. "Collusion Could Work." *Foreign Policy* 14 (Spring 1974): 57-68.

_____. *The Community of Oil Exporting Countries*. Ithaca, N.Y.: Cornell University Press, 1972.

_____. "Cooperation Among Oil Exporting Countries with a Special Reference to Arab Countries: A Political Economy Analysis." *International Organization* 28 (Winter 1974): 1-30.

_____. *A Financial Analysis of Middle Eastern Oil Concessions, 1901-1965*. New York: Praeger, 1966.

_____. *The International Politics of Natural Resources*. Ithaca, N.Y.: Cornell University Press, 1976.

_____; Cleland, S.; and Seymour, I., eds. *Continuity and Change in the World Oil Industry*. Beirut: Middle East Resource and Publication Center, 1970.

Mikesell, Raymond F. "Nonfuel Minerals: "U.S. Investment Policies Abroad." Washington Papers, Vol. 3, No. 23. Beverly Hills: Sage Publications, 1975.

Mingst, Karen A. "Cooperation or Illusion: An Examination of the Intergovernmental Council of Copper Exporting Countries." *International Organization* 30 (Spring 1976): 263-87.

_____. "Regional Sectorial Economic Integration: The Case of OAPEC." *Journal of Common Market Studies* 16 (December 1977): 91-113.

Morse, Edward L. "Crisis Diplomacy, Interdependence and the Politics of International Economic Relations." *World Politics* 24 (Spring 1972), Supplement, 123-50.

Mytelka, Lynn K. "Fiscal Politics and Regional Redistribution." *Journal of Conflict Resolution* 19 (March 1975): 138-60.

Nichols, Albert L., and Zeckhauser, Richard J. "Stockpiling Strategies and Cartel Prices." *Bell Journal of Economics* 8 (Spring 1977); 66-96.

Noll, Roger D. *Reforming Regulation*. Washington, D.C.: The Brookings Institution, 1971.

Noreng, Øystein. "Friends or Fellow Travelers? The Relationship of Non-OPEC Exporters with OPEC." *Journal of Energy and Development* 4 (Spring 1979): 313-35.

_____ . *World Oil Politics in the 1980's: Patterns of Cooperation*. New York: McGraw-Hill, 1978.

Nye, Joseph S. "Comparing Common Markets." *International Organization* 24 (Autumn 1970): 796-835.

_____ . *Peace in Parts*. Boston: Little, Brown and Company, 1971.

_____ , and Keohane, Robert O. "Transnational Relations and World Politics." *International Organization* 25 (Summer 1971): 329-49.

Organization of Arab Petroleum Exporting Countries. *A Brief Report on the Activities and Achievements of the Organization, 1968-73*. Kuwait: Al-Qabab, 1974.

_____ . *Secretary General's First Annual Report Presented to the 13th Ordinary Meeting of the Council of Ministers*. Kuwait: OAPEC, 1974.

_____ . *Secretary General's Second Annual Report*. Kuwait: OAPEC, 1975.

_____ . *Secretary General's Third Annual Report*. Kuwait: OAPEC, 1976.

_____ . *Secretary General's Fourth Annual Report*. Kuwait: OAPEC, 1977.

_____ . *Secretary General's Fifth Annual Report*. Kuwait: OAPEC, 1978.

_____ . *Third Annual Statistical Report 1974-1975*. Kuwait: OAPEC, 1976.

_____ . *Fourth Annual Statistical Report 1975-1976*. Kuwait: OAPEC, 1977.

_____ . *Fifth Annual Statistical Report 1976-1977*. Kuwait: OAPEC, 1978.

_____ . *Opportunities for Cooperation with the Arab World*. London: Bentley Brothers, 1974.

_____ . *Sources and Problems of Arab Development*. Kuwait, 1980.

Organization of Petroleum Exporting Countries. *Selected Documents—1972*. Vienna: OPEC, 1973.

_____ . *Annual Report, 1978*. Vienna: OPEC, 1978.

Ortiz, Rene G. "International Relations: OPEC as a Moderating Political Force." *OPEC Review* 4 (Summer 1980): 1-7.

_____ . "The World Energy Outlook in the 1980's and the Role of OPEC." *Journal of Energy and Development* 4 (Spring 1979): 197-211.

Otaiba, Mana Saeed al-. *OPEC and the Petroleum Industry*. London: Croom Helm, 1975.

Oweiss, Ibrahim M. "Strategies for Arab Economic Development." *Journal of Energy and Development* 3 (Autumn 1977): 103-14.

Penrose, Edith. "The International Oil Industry in the Middle East." *Middle East Economic Survey Supplement*, 2 August 1968.

Piccini, Raymond. "On the Effects of Energy and Conservation on OPEC Pricing." *Journal of Energy and Development* 3 (Autumn 1977): 190-92.

Pindyck, Robert S. "Cartel Pricing and the Structure of the World Bauxite Market." *The Bell Journal of Economics* 8 (Autumn 1977).

_____ . "Gains to Producers from the Cartelization of Exhaustible Resources." 1976. Mimeographed.

————. "Some Long-Term Problems in OPEC Oil Pricing." *Journal of Energy and; Development* 4 (Spring 1979): 259-72.

Poulson, Barry W. and Wallace, Myles. "Regional Integration in the Middle East: The Evidence for Trade and Capital Flows." *The Middle East Journal* 33 (Autumn 1979): 464-78.

Puchala, Donald J. "Domestic Politics and Regional Harmonization in the European Communities." *World Politics* 27 (July 1975): 496-520.

————. "International Transactions and Regional Integration." *International Organization* 24 (Autumn 1970): 732-63.

Rand, Christopher T. *Making Democracy Safe for Oil*. Boston: Little, Brown and Company, 1975.

Rodinson, Maxime. *Islam and Capitalism*. Translated by Brian Pearce. Austin: University of Texas Press, 1978.

Rosecrance, Richard; Alexandroff, Alan; Koehler, Wallace; Kroll, John; Lacquer, Shlomit; and Stocker, John. "Whither Interdependence?" *International Organization* 31 (Summer 1977): 425-72.

Rouhani, Fuad. *A History of O.P.E.C.* New York: Praeger, 1971.

Rushdi, Mahmoud. "OAPEC's Role in Promoting Regional Cooperation Among Member States." Paper presented at the Tenth Arab Petroleum Congress, Tripoli. No date.

Russett, Bruce M. *International Regions and the International System*. Chicago: Rand McNally, 1967.

Rustow, Dankwart A., and Mugno, J. F. *OPEC: Success and Prospects*. New York: New York University Press, 1976.

Sadat, Anwar el-. *In Search of Identity*. New York: Harper & Row, 1977.

Saikal, Amin. *The Rise and Fall of the Shah*. Princeton: Princeton University Press, 1980.

Sampson, Anthony. *The Seven Sisters*. New York: Viking Press, 1975.

Savage, I. Richard and Deutsch, Karl W. "A Statistical Model of the Gross Analysis of Transaction Flows." *Econometrica* 28 (July 1960): 551-72.

Sayegh, Fayez A. *Arab Unity: Hope and Fulfillment*. New York: Devin Adair, 1958.

Sayigh, Yusif A. "Problems and Prospects of Development in the Arabian Peninsula." *Journal of Middle East Studies* 2 (January 1971): 40-58.

Scherer, Frederic M. *Industrial Market Structure and Economic Performance*. Chicago: Rand McNally, 1970.

Schmalensee, Richard. "Resource Exploitation Theory and the Behavior of the Oil Cartel." *European Economic Review* 7 (April 1976): 257-79.

Schmitter, Phillippe C. "A Revised Theory of International Integration." *International Organization* 24 (Autumn 1970); 836-68.

————. "Autonomy or Dependence as Regional Integration Outcomes: Central America." Institute of International Studies Research Series No. 17. Berkeley: University of California Press, 1972.

Schneider, William. *Food, Foreign Policy and Raw Materials Cartels*. New York: Crane, Russak, 1976.

Schultz, Lawrence E. "OPEC: Deceased Theories Survived by the Patient: A Study of Organizational Response to Interdependence." 1977. Mimeographed.

Segal, Aaron. "The Integration of Developing Countries: Some Thoughts on Africa and Central America." *Journal of Common Market Studies* 2 (June 1967): 263-82.

Serafy, Salah el-. "The Oil Price Revolution of 1973-74." *Journal of Energy and Development* 4 (Spring 1979): 273-90.

Shaath, Nabeel. "High Level Palestinian Manpower." Reprinted from *Journal of Palestine Studies* 1 (Winter 1972), New York: Arab Information Center, 1972.

Sherbiny, Naiem A., and Tessler, Mark A., eds. *Arab Oil: Impact on the Arab Countries and Global Implications.* New York: Praeger, 1976.

Shihata, Ibrahim. *The Case for the Arab Oil Embargo.* Beirut: Institute for Palestine Studies, 1975.

_____. "OPEC Aid, the OPEC Fund, and Cooperation with Commercial Development Finance Sources." *Journal of Energy and Development* 4 (Spring 1979): 291-303.

_____, and Mabro, Robert. *The OPEC Aid Record.* London: OPEC Special Fund, 1978.

Stevens, G. and H. "The First Arab Petroleum Congress." *The World Today* 15 (June 1959): 246-53.

Stocking, George W. *Middle East Oil: A Study in Political and Economic Controversy.* London: Allen Lane, 1970.

_____, and Watkins, Myron W. *Cartels in Action.* New York: The Twentieth Century Fund, 1947.

Stork, Joe. *Middle East Oil and the Energy Crisis.* New York: Monthly Review Press, 1975.

Taher, Abdulhady H. "The Middle East Oil and Gas Policy." *Journal of Energy and Development* 3 (Spring 1978): 260-69.

Tanzer, Michael. *The Energy Crisis: World Struggle for Power and Wealth.* New York: Monthly Review Press, 1974.

_____. *The Political Economy of International Oil and the Underdeveloped Countries.* Boston: Beacon Press, 1969.

Tétreault, Mary Ann. "Measuring Interdependence." *International Organization* 34 (Summer 1980): 429-43.

Tharp, Paul A., Jr. "Transnational Enterprises and International Regulation: A Survey of Various Approaches in International Organizations." *International Organization* 30 (Winter 1976): 47-73.

Thoman, R. E. "Iraq and the Persian Gulf Region." *Current History* 62 (January 1973): 21-25, 37-38.

_____. "The Persian Gulf Region." *Current History* 60 (January 1971): 38-45, 50.

Todaro, Michael. *Economic Development in the Third World.* New York: Longman and Sons, 1977.

Tollison, Roger, and Willett, Thomas. "International Integration and the Interdependence of Economic Variables." *International Organization* 27 (Spring 1973): 255-71.

Tomeh, George J. "Arab Politics and Priorities in Economic Cooperation with Western Europe." In *Euro-Arab Cooperation,* edited by Edmond Volker. Leydon: A. W. Sijthoff, 1976.

————. "Interdependence, International Cooperation and Natural Resources." Discussion paper presented to the Kuwait session of the Wisconsin Seminar on Natural Resource Policies in Relation to Economic Development and International Cooperation." February 1978.

————. "OAPEC, Its Affiliated Firms and Arab Regional Organizations." Lecture to the British House of Commons, 22 February 1977.

————. "OAPEC: Its Growing Role in Arab and World Affairs." *Journal of Energy and Development* 1 (Autumn 1977): 26-36.

Tugendhat, Christopher. "Political Approaches to the World Oil Problem." *Harvard Business Review*, January-February 1976, pp. 45-55.

U.S. Congress, Senate, Committee on Foreign Relations, Subcommittee on Multinational Corporations. *Multinational Corporations and United States Foreign Policy, Hearings*, 10 parts. 93rd Congress, First Session. Washington, D.C.: G.P.O., 1975.

Vernon, Raymond. *Multinational Enterprise and National Security*. Adelphi Papers, No. 74. London: Institute for Strategic Studies, 1972.

————, ed. *The Oil Crisis*. New York: W. W. Norton, 1976.

Victor, Ray. *The Kingdom of Oil*. New York: Charles Scribner's Sons, 1974.

Vinogradov, Amal. "The 1920 Revolt in Iraq Reconsidered: The Role of Tribes in National Politics." *International Journal of Middle East Studies* 3 (April 1972): 123-39.

Von der Mehden, Fred. *Communalism, Wealth and Income in Afro-Asia*. Rice University Program of Development Studies, August 1977.

Wattari, Abdul Aziz al-. "Manpower Use and Requirements in OAPEC States." Paper presented at the Seminar on Development Through Cooperation Between Scandanavian and OAPEC Countries, Oslo, Norway, 27-29 September 1978.

Weisberg, Richard C. *The Politics of Crude Oil Pricing in the Middle East, 1970-1975*. Research Series 31. Berkeley: University of California Press, 1977.

Williams, Maurice J. "The Aid Programs of the OPEC Countries." *Foreign Affairs* 54 (January 1976): 308-24.

Wilson, James Q. *Political Organizations*. New York: Basic Books, 1973.

Wren, Christopher S. "Peace Hasn't Been Easy for Egypt's Army." *The New York Times*, 23 September 1979; p. 2-E.

Yamani, Ahmed Zaki. "Energy Outlook: The Year 2000." *Journal of Energy and Development* 5 (Autumn 1979): 1-8.

Young, Crawford. *The Politics of Cultural Pluralism*. Madison: University of Wisconsin Press, 1976.

Young, Oran, M. "Political Discontinuities in the International System." *World Politics* 20 (April 1968): 369-92.

Zahlan, A. B. *Science and Science Policy in the Arab World*. New York: St. Martin's Press, 1980.

Index

Abu Dhabi, 41, 49, 141, 146, 149, 152, 153
Adelman, M. F., 130, 131
Alexandria protocol, 37
Algeria, 28n, 44, 48, 49, 53, 72, 76, 80, 98, 114, 129, 138, 139, 140, 141, 142, 143, 144, 145, 146, 147, 150, 153, 154, 155, 157, 160n, 164n, 170, 171
Amin, Mahmoud, 69, 84n
Anbari, Abdul Amir al-, 149
Arab boycott, 55n, 56n
Arab cold war, 37
Arab community, 7, 11, 16, 22, 37, 88, 89, 91, 92, 169
  Institutional, 88, 89, 104, 105, 119, 120
  Structural, 89, 90, 103, 104, 119, 120
Arab-Israeli wars, 14, 42, 53, 88, 169
Arab League. See League of Arab States
Arab nationalism, 7, 11, 12, 13, 14, 15, 30n, 31n, 70, 98
Arab/OPEC foreign aid, 112
Arab petroleum congress(es), 39, 42, 51, 54n
Arab scientific activity, 78, 79, 80
Arabian Gulf, 54n, 74n, 86n, 100, 102
Arabian Light, 149, 154, 156, 157, 170
Arabian Peninsula, 14, 30n, 31n, 91, 96, 98
Arabic language, 12
Arabization, 56n, 80, 81
Arafat, Yassir, 98
ARAMCO (Arabian American Oil Company), 15, 45, 68, 149, 151
Attiga, Ali A., 70, 72, 84n, 85n, 87n, 169

Bahrain, 28n, 49, 72, 74, 75, 76, 84n, 85n, 87n, 140, 142, 144, 145, 169
Buffer system(s), 68, 69

Canada, 126, 135
Capital absorption capacity, 144, 145
Cartel, 5, 6, 7, 23, 24, 25, 26, 27, 28n, 34n, 35n, 36n, 63, 106, 125, 126, 128, 130, 132, 135, 136, 137, 157, 158, 158n, 159n, 160n, 161n, 163n, 165n
Cartel cheating, 63, 126
Ceiling price, 150, 151, 152, 154, 157, 170
Collective goods, 18, 19
Colonialism, 11
Committee of Arab Oil Experts, 38, 51, 65, 66
Common market(s), 17, 19, 20, 21, 27n, 33n, 63, 104
Consensus decisionmaking, 59
Contract premiums, 150
Crude carriers, very large (VLCCs), 74

Demand, 135, 136, 148
Dependency, 22, 33n, 34n, 78, 87n, 92, 96, 118, 169
Development funds and banks, 44, 112
Differentials, 40, 150, 151, 152, 154, 158, 163n, 164n, 165n, 170
Distribution, 19, 132, 133, 134, 136, 137, 146, 158
Doha Conference (1976), 135, 156, 163n
Downstream investment, 11, 29n, 30n, 57n, 67, 72, 74, 76, 79, 84n, 86n, 105, 106
Dubai, 49, 106, 107, 108, 109, 141

Economic commission, 126
Economic development, 6, 7, 8, 9, 10, 11, 18, 23, 28n, 29n, 35n, 44, 45, 50, 55n, 57n, 68, 76, 77, 81, 83n, 89, 91, 97, 99, 103, 106, 168, 171
Ecuador, 142, 143, 148
Egypt, 8, 13, 14, 15, 28n, 29n, 30n, 32n, 34n, 43, 45, 50, 54n, 56n, 66, 72, 88, 91, 96, 99, 100, 101, 102, 103, 104, 114, 120, 140, 144, 145, 150, 151, 169
Excess production capacity, 143
Exporter fringe, 125

Fertile Crescent, 12, 14, 30n, 31n, 91, 96

Floor price, 154, 156, 170
Foreign aid, 113, 114, 115
France, 12, 13, 25, 44, 79, 87n, 96
Free rider(s), 126, 131, 158

Gabon, 142, 143

Ibn Sa'ud, 'Abdul 'Aziz, 13, 15
Independent oil companies, 160n
Indonesia, 42, 45, 138, 139 141, 142, 143,
    146, 147, 153, 155, 163n
Infrastructure, 8, 19, 114
Integration. *See* Regional integration
Interdependence, 7, 17, 18, 20, 21, 22,
    33n, 34n, 38, 50, 51, 57n, 89, 90, 96, 99,
    101, 102 103
International Energy Agency (IEA), 131
Iran, 29n, 38, 39, 41, 42, 43, 45, 48, 56n,
    70, 90, 91, 106, 107, 127, 138, 139, 141,
    142, 143, 146, 147, 148, 151, 152, 153,
    155, 156, 157, 158, 160n, 163n, 164n,
    166n, 170
  Shah of Iran, 6, 7, 28n, 83n, 148, 159n,
    171
Iranian revolution, 78, 89, 90, 109, 131,
    142, 148, 150, 152
Iraq, 12, 13, 15, 28n, 30n, 39, 41, 42, 43,
    44, 46, 47, 48, 49, 50, 53, 54n, 58, 72, 88,
    91, 97, 99, 103, 120, 130, 138, 139, 140,
    141, 142, 143, 144, 145, 146, 147, 148,
    149, 150, 151, 152, 153, 155, 156, 163n,
    164n, 165n, 170
Islam, 7, 12, 31n, 90
Israel, 5, 6, 14, 15, 30n, 31n, 32n, 38, 43,
    51, 53, 54, 54n, 56n, 65, 88, 91, 97, 99,
    103, 120, 130, 150

Jordan, 8, 15, 32n, 45, 75, 80, 98, 99, 100,
    102, 103
Judicial Board protocol, 109, 114, 119
June (1967) war, 43

Khalifa bin Salman bin Muhammad, 74,
    85n
Kuwait, 8, 9, 13, 28n, 29n, 35n, 39, 42, 43,
    44, 45, 46, 48, 54n, 57n, 66, 72, 77, 79,
    80, 83n, 98, 114, 117, 130, 138, 139, 141,

142, 143, 144, 145, 146, 147, 148, 149,
    150, 151, 152, 153, 155, 156, 159n,
    160n, 163n, 165n, 170

Labor migration, 8, 100, 101, 102, 103,
    104
League of Arab States, 14, 22, 29n, 31n,
    37, 38, 39, 48, 49, 50, 51, 52, 53, 54n,
    57n, 58, 59, 60, 61, 62, 63, 64, 65, 66, 69,
    81, 82n, 89, 90, 91, 92, 103, 105, 113,
    117, 118, 119, 120
Leapfrogging, 42, 129, 151, 152, 160n
Lebanaon, 12, 28n, 90, 99, 103
Libya, 15, 28n, 29n, 35n, 41, 42, 43, 44, 45,
    46, 48, 49, 50, 52, 53, 63, 66, 72, 75, 86n,
    98, 100, 102, 103 119, 127, 128, 129,
    130, 138, 139, 141, 142, 143, 144, 145,
    146, 147, 149, 150, 151, 153, 155, 159n,
    160n, 163n, 164n, 165n, 170
Logrolling, 133, 134

Marker crude, 125, 149, 150, 151, 152,
    154, 156, 157, 170
Market power, 38, 140, 141, 156
Market shares (petroleum), 130, 136, 144
Mexico, 34n, 125, 131
Middle East Economic Survey (MEES),
    5, 28n, 29n, 32n, 56n, 57n, 66, 75, 83n
Most favored seller clause, 150, 151
Multinational oil companies. *See* Oil
    companies

Nasser, Gamal Abdal, 8, 31n, 45, 54n
National sovereignty, 63, 76
National tanker fleets, 70, 74, 77, 107,
    108, 109, 117
Nationalistic investment, 67, 118
Nationalization of oil companies, 15, 47
Nigeria, 141, 142, 143, 146, 147, 148, 150,
    153, 155
Norway, 125, 131, 135

OAPEC (Organization of Arab Petro-
    leum Exporting Countries), 5, 6, 7, 8, 9,
    10, 11, 22, 23, 25, 27n, 28n, 29n, 31n,
    32n, 46, 47, 48, 49, 50, 52, 54n, 56n,
    57n, 58, 59, 60, 61, 62, 63, 64, 65, 66, 67,

68, 69, 70, 71, 72, 73, 74, 75, 76, 78, 79, 80, 81, 82, 83n, 84n, 85n, 88, 89, 90, 91, 92, 104, 105, 106, 108, 109, 110, 112, 113, 114, 115, 116, 117, 118, 119, 120, 130, 137, 139, 140, 142, 143, 144, 145, 146, 147, 150, 156, 157, 158, 168, 169, 170, 171, 172

OAPEC Agreement, 29n, 46, 47, 48, 49, 50, 53, 56n, 57n, 58, 60, 61, 62, 64, 67, 71, 75, 76, 81, 82n, 83n, 84n, 85n, 118, 119

OAPEC Council of Ministers, 5, 23, 48, 49, 62, 65, 66, 67, 70, 71, 73, 75

OAPEC, information exchange within, 67, 68

OAPEC joint venture companies, 70, 71, 73, 75, 76, 78, 80, 81, 82, 85n, 88, 105, 109, 114, 117, 118, 119, 157, 168, 169, 171, 172

General Assembly, 73

OAPEC joint ventures, 70, 74, 79, 84n, 106, 107, 108, 171

  Arab Drilling and Workover Company (ADWC), 75, 78, 84n, 86n

  Arab Engineering Consulting Company (AECC), 115, 168

  Arab Maritime Petroleum Transport Company (AMPTC), 67, 72, 74, 76, 79, 85n, 86n, 87n, 105, 106, 107, 108, 109, 115, 169, 170

  Arab Petroleum Investments Corporation (APICORP), 72, 74, 75, 77, 78, 84n, 85n, 86n, 114, 115

  Arab Petroleum Services Company (APSC), 71, 72, 75, 78, 84n, 85n, 114, 115

  Arab Petroleum Training Institute (APTI), 75, 78, 79, 80, 82, 169, 172

  Arab Shipbuilding and Repair Yard (ASRY), 67, 72, 74, 76, 79, 81, 85n, 86n, 87n, 105, 107, 108, 109

OAPEC Judicial Board, 52, 61, 62, 65, 75, 76, 81, 82, 118, 119, 120, 169, 170

OAPEC Legal Committee on Maritime Transport, 70, 71

OAPEC, membership in, 46, 47, 58, 168

OAPEC resolutions, 57n, 61

OAPEC Secretariat, 60, 67, 69, 71, 77, 80, 112, 172

OAPEC Standing Committee on Refining, 70

Occupied territories, 14

October (1973) war, 5

Oil and Gas Journal, 55n, 154

Oil companies, 5, 40, 42, 46, 47, 63, 65, 66, 67, 68, 69, 75, 77, 78, 80, 106, 128, 129, 130, 131, 132, 133, 137, 157, 158, 165n, 168

Oil conservation, 128, 132, 136, 141

Oil-consuming countries, 133

Oil crisis, 5, 17, 18, 27n, 28n, 35n, 55n, 148, 149

Oil embargo
  of 1967, 44, 51
  of 1973, 48, 53, 57n, 129, 130, 154, 157, 171

Oil embargoes, 5, 27n, 28n, 43, 44, 45, 52, 56n, 61, 126, 131, 146, 157, 160n, 168

Oil glut, 42, 109, 125, 127, 139

Oil income, 28n, 130, 132, 136, 144, 145, 147

Oil prices, 5, 6, 7, 10, 23, 24, 25, 26, 27, 35n, 36n, 39, 40, 43, 44, 45, 46, 52, 53, 55n, 56n, 57n, 66, 112, 113, 125, 126, 127, 128, 129, 130, 131, 132, 133, 135, 136, 139, 140, 141, 142, 143, 144, 146, 159n, 160n, 161n, 163n, 164n, 165n, 166n, 170, 171, 172

Oil production, 15, 27n, 28n, 46, 65, 66, 126, 127, 128, 130, 132, 135, 137, 139, 140, 141, 142, 143, 144, 146, 148, 151, 152, 156, 157, 158, 163n, 164n, 165n

  Cuts, 5, 46, 129, 130, 136, 139, 140, 141, 149, 157, 158, 170

  Prorationing, 26, 70, 126, 127, 135, 157, 158

Oil reserves, 28n, 117, 137, 138, 140, 141, 144, 145, 147, 158, 163n, 165n

Oil revenues, 8, 27n, 29n, 44, 59, 99, 110, 111

Oil shortage, 7, 36n, 149, 152, 161n

Oil supply, 141, 142, 143, 148, 157, 171

Oil tankers, 11, 128

Oil weapon, 43, 44, 45, 51, 56n, 57n, 65, 130, 132

OPEC (Organization of Petroleum Exporting Countries), 5, 6, 7, 23, 25, 26, 27, 27n, 28n, 29n, 35n, 36n, 39, 40, 42, 46, 48, 50, 51, 52, 54n, 55n, 58, 59, 60, 61, 62, 63, 64, 66, 67, 69, 70, 71, 72, 81, 83n, 84n, 101, 103, 105, 106, 118, 125, 126, 127, 128, 129, 130, 131, 132, 133, 134, 135, 136, 137, 138, 139, 140, 141, 142, 143, 144, 146, 147, 148, 150, 151, 152, 153, 154, 155, 156, 157, 158, 159n, 160n, 161n, 163n, 164n, 165n, 168, 170, 171, 172

OPEC Conference meetings, 40, 45, 55n, 62, 69, 126, 149, 150, 151, 154, 156, 165n, 170

OPEC Long Term Strategy Committee, 127

OPEC resolutions, 55n, 60, 63

Ortiz, Rene G., 148, 161n,

Otaiba, Mana Saeed al-, 148

Pact of the League of Arab States, 60

Palestine, 12, 14, 31n, 54n, 61

Palestine Liberation Organization (PLO), 98

Palestinians, 31n, 97, 98, 99, 102, 103

Pan-Arabism, 12, 88, 120

Partial (Arab) unification, 15

Participation. See Nationalization of oil companies

Petrochemicals, 9, 11, 74, 79, 86n, 92

Policy arena(s), 132, 133, 134, 135, 136, 157, 161n

Posted prices, 23, 39, 40, 54n, 55n

Price range, 149

Production program, first and second, 127

Production shares, 36n, 127, 135

Public interest arena, 134, 135

Qadhafi, Mu'Ammar al-, 66

Qatar, 8, 28n, 29n, 41, 46, 49, 72, 86n, 138, 139, 141, 142, 143, 144, 145, 146, 148, 151, 152, 153, 163n

Redistribution, 18, 19, 21, 22, 33n, 34n, 63, 70, 113, 114, 132, 133, 134, 135, 163n

Refining, 10, 140

Regional integration, 16, 17, 18, 20, 22, 28n, 32n, 33n, 34n, 38, 88, 89, 120

Royalty expensing, 40, 42, 45, 55n, 56n, 159n

Sadat, Anwar el, 8, 15, 32n

Sa'dawi, Suhail, 48, 66

Saudi Arabia, 6, 8, 13, 14, 15, 28n, 32n, 39, 41, 42, 43, 44, 45, 46, 48, 50, 53, 58, 62, 66, 72, 127, 131, 135, 136, 138, 139, 140 141, 142, 143, 144, 145, 146, 147, 148, 149, 150 151, 152, 153, 154, 155, 156, 157, 160n, 163n, 164n, 170, 171

Schlesinger, James, 148

Schuler, George, 128, 129

Solar energy, 9

Soviet Union, 91, 125

Spot oil market, 131, 149, 150, 151, 152

Stevens, Georgiana, 76

Suez Canal, 13, 43, 44, 45

Suez crisis, 43

Surcharges, 151

Syria, 12, 13, 14, 15, 28n, 31n, 32n, 43n, 44, 50, 56n, 57n, 72, 80, 88, 99, 114, 140, 144, 145, 169

Tacit collusion, 158n

Tapline (Trans Arabian Pipeline), 45

Tariki, Abdullah, 39, 109

Technology independence, 10, 11, 78, 82, 169

Technology transfer, 6, 7, 78, 79, 80, 81, 82, 118, 168, 171

Teheran Agreement, 129

Tomeh, George G., 55n, 57n, 83n, 84n, 87n

Training programs, 114, 118

Turn-key plants, 9, 10

Two-tier pricing, 151, 156, 163n, 164n

United Arab Emirates (UAE), 28n, 29n, 34n, 72, 114, 138, 139, 141, 142, 143, 144, 145, 146, 147, 148, 151, 152, 155

United Arab Republic (UAR), 15, 31n, 44, 45, 48, 52, 57n, 88
United Kingdom, 12, 13, 25, 30n, 43, 54n, 74, 79, 88, 131
United States, 6, 23, 24, 28n, 43, 78, 91, 92, 96, 131, 135, 148, 150

Venezuela, 38, 39, 42, 43, 45, 130, 138, 139, 141, 142, 143, 146, 147, 151, 153, 155, 163n, 164n, 165n

Wall Street Journal, 6, 28n
Windfall profits, 131
World oil, 6, 36n, 70

Yamani, Ahmad Zaki, 43, 47, 56n, 57n, 65, 83n, 127, 152
Yemen, 15, 45, 102, 103

About The Author
Mary Ann Tétreault is Assistant Professor of Political Science at Old Dominion University in Norfolk, Virginia. She specializes in the study of energy policy and international political economy.